ENCYCLOPAEDIA OF PETROLEUM SCIENCE AND ENGINEERING

ENCYCLOPAEDIA OF PETROLEUM SCIENCE AND ENGINEERING

Vol–Six

(In Eight Volume)

Surveying, Geophysical Field System, Seismic Stratigraphy and Log Analysis of Subsurface Geology

S. L. Sah

KALPAZ PUBLICATIONS

DELHI - 110052

ENCYCLOPAEDIA OF PETROLEUM SCIENCE AND ENGINEERING

ISBN : 81-7835-102-1 (Set)
ISBN : 81-7835-298-6 (Vol. 6)

Published in 2004 in India by
Kalpaz Publications,
C-30, Satyawati Nagar,
Delhi-110 052
Email : kalpaz@hotmail.com

Lasertype Setting by: Tact Computer Graphics, Delhi
Printed at: Mehra Offset Press, Delhi

Dedicated to the geophysicists, geologists, engineers, scientists, planners, universities, organizations, teachers, students, and other working in different disciplines of upstream techniques.

"Oil Is Found First In The Mind"

Wallace Pratt

Contents

Preface

The hydrocarbon industry is today the single largest provider of commercial energy across the world with petroleum contributing the lion's share. Despite successes in the development and utilisation of part-substitutes, any viable alternative that can completely replace petroleum fuels still remains a distant possibility. Petroleum will, therefore, continue to dominate the global energy mix at least for some more time. For a fast developing and vast country like India even a 1ow consumption growth rate translates into large volumes as is evident from its oil demand projections over the next five years. At an envisaged annual growth rate of 6 to 7 per cent, oil consumption in the country is estimated to peak at 134.5 million tonnes by 2006-07 from the current level of nearly 100 million tonnes. Contrasting this with our domestic crude oil production hovering around 30 per cent of demand. It is clear that our upstream oil industry has immense scope for growth and offers unlimited opportunities for exploration and production companies to participate in this area. Petroleum is a technology-intensive industry and needs strong R&D back-up for upgradation of inuse technologies and development and adoption of newer technologies to meet changing parameters of processes, product quality and environmental compliance. India is fortunate to have been able to develop a high level of R&D expertise and excellent infrastructure to provide technology solutions to upstream areas.

In the nineties, a significant change occured in hydrocarbon industry in India. In line with the merger and acquisition that took place globally, in India too, major restructuring of the national oil companies took place in the form of merger of smaller companies with larger companies. The upstream sector was also considerably changed with the opening of exploration for national and international private investors and a number of overseas ventures of ONGC. Today, the Indian Petroleum Industry is contributing to the energy security through excellence in science and technology and improving the quality of life while being environmentally friendly and responsive to the preservation of national cultural heritage. Petroleum industry is knowledge and technology intensive. Since inception, the leadership in Indian Petroleum Industry has been alive to the potential of research and development, design engineering and technology integration for efficient solutions and development of new generation of technologies to keep the industry fresh, vibrant, innovative and competitive. Deliberate and focused R&D in hydrocarbon sector with greater state sponsorship is a well-recognised key result area to achieve the necessary self- reliance in technology towards meeting the objective of energy security.

Chapter–one of this encyclopaedia gives about the surveying. Surveying may be defined as the art of making measurements of relative positions of natural and man-made features on the earth's surface and the presentation of this information either graphically or numerically. It is divided primarily into geodetic surveying and plane surveying. In geodetic surveying large areas of the earth's surface are envolved and the curvature of the earth must be taken into account. In plane surveying relatively small areas are under consideration, and it is taken that the earth's surface is flat, i.e., it gives a horizontal plane. Surveys are made for a variety of purposes such as the determination of areas, the fixing of boundary lines and the plotting of maps. Furthermore, engineering constructions, such as water works, railroads, mines, bridges, and building, all require surveys. The study of terrain features and conditions by aerial photography has developed to expedite a variety of engineering work. Now accurate position can be found By Global Positioning System (GPS). By using this system, the user can determine his position in three dimensions as well as the system time. With GPS, it is possible to achieve an accuracy of 5 meteres or better in satellite position fixes.

Chapter– two of this encyclopaedia gives about geophysical field system. The most important methods of geophysical exploration represent a primary tool for investigation of the subsurface and are applicable to a very wide range of problems. Geophysical field system is used for recording the seismic data. Today's objective is to search geological features at all the depths that have remained unresolved because of small size, geological complexity or poor data. Improved seismic methods of data acquisition and recording are the key to successful exploration of the more demanding prospecting requirements. The present trend in seismic exploration is towards improved sources, geophones, field procedures, data recording and data processing. New data acquisition systems capable of simultaneously recording more than 1000 data channels have brought about the possibility of point-source point-detector mode of data acquisition, which minimizes the smear of ground arrays and reduces the blurring of data. Seismic recording equipment is a complex electronic system designed by highly specialised electronics engineers. We can anticipate new technological developments in all aspects of geophysical exploration in acquisition, processing and interpretation. But in the real world, socio-political, economic and environmental factors can and do play an important role in the developments or non-development of future technology.

Chapter–three of this encyclopaedia gives about seismic stratigraphy. In exploration of unknown areas, determining the nature of the rocks present is often the difficult aspect. Most of the properties of rocks that can be measured at a distance are not distinctive enough to identify the rock unambiguously. Coventionally, seismic data are used to determine aspects of the structure. Stratigraphic pictures are inferred from the structure, the nature of rocks exposed for examination in the surrounding area, and regional concepts. Seismic reflections carry information about the interfaces between rock types in the earth. Lateral changes in rock type are in general rule, although usually rock type changes occur gradually. Inferring stratigraphic changes and where they occur based on characteristics of seismic data is an objective of seismic stratigraphy. The distinction of seismic stratigraphy is that it goes beyond merely mapping structure and search for more direct seismic evidence as to the nature of rocks the and the fluid contained within their pore spaces. Stratigraphy thus embraces most of what one might consider the geology of sedimentary rocks.

Chapter-four of this encyclopaedia gives about log analysis of subsurface geology. The logs give useful insights on subsurface geology. In recent years, there has been a steady evolution within log analysis from methods grounded firmly in classical reservoir engineering theory to a more integrated approach which incorporates geological concepts and interpretations. The subtleties of stratigraphic trap and the complexities of tertiary oil recovery projects have spurred the design and application of appropriate geological models. The revolutionary advances in seismic data processing which led to seismic stratigraphy often require simulation models based on logs to extract their full geological meaning. At the same time, a new generation of logging tools has been introduced which measure properties that reveal patterns of deposition and diagenesis. However, the full potential of modern logs as a medium for enhanced interpretations of subsurface geology draws ultimately on the fusion of an intelligent understanding of several areas. These include a familiarity with the principles of logging measurements, a knowledge of reservoir engineering fundamentals, an informal applications of simple mathematics and computer processing, and a sound background in geology.

The aim of this encyclopaedia is to make the interconnection among the different disciplines of upstream techniques like add surveying, geophysical field systems, seismic stratigraphy, and log analysis of subsurface geology.

At the end of this encyclopaedia six appendixes have been included. These appendixes will give information to readers about the upstream techniques.

This encyclopaedia will help to promote understanding and communication among users. It is suitable for geophysicists, geologists, scientists, engineers, planners, universities, organisations, teachers, students, and other working in different disciplines of upstream.

The author will be grateful for comments and criticism which might help to improve the later edition of this encyclopaedia.

Some of the material of this encyclopaedia has been taken from the books and the papers published in different journals. I am thankful to all of them who have contributed to the development of this encyclopaedia.

S. L. Sah

One
Surveying

Introduction

If knowledge is power, then inspiration to learn is the power of our potential. This potential power moves us from the systematic to the spontaneous, and from ideas to results.

Surveying has as its purpose the determination of the dimensions and contour of any part of the earth. Establishing the relative position on points at or near the earth's surface requires both linear and angular measurements. Surveying operations are confined to land, water and extra terrestrial space.

The development of measuring techniques and units of measure parallel the development of civilisation itself, for there has always been a need for measurement. Primitive man used night and day and the cycle of the seasons as measures of time.

Principles and practices of surveying have been connected with engineering. The art of measuring is fundamental in many engineering operation in many fields. It forms the base upon which surveying rests.

A knowledge of surveying is advantageous in many phases of engineering and provides excellent training in the ability to visualize, to think logically, and to pursue the "engineering approach".

The application of surveying requires skill as well as knowledge of mathematics, physics, and to some extent, astronomy. Today, surveying instruments and methods are used at some point in nearly every engineering undertaking. Often the basic principles of surveying find application in all of the stages of a project from initial planning to final completion.

The study of terrain features and conditions by aerial photography has developed to expedite a varity of engineering work. Aerial cameras are mounted in the aircraft so that in flight the axis of the lens is vertical or at a known angle of obliquity. The exposure is made through an opening in the plane's fusilage. Photographs taken from the air may be classified as vertical, oblique or composite.

Now accurate position can be found by Global Positioning System (GPS). By using this system, the user can determine his position in three dimension as well as the system time. With GPS it is possible to achieve an accuracy of 5 metres or better in satellite position fixes. This system consists of a network of 24 satellites which are orbiting in three different planes.

Detailed Encyclopaedia

This encyclopaedia is arranged in alphabetical order. The detailed encyclopaedia is given below:–

Accretion–It results in either alluvium or reliction and the boundary of property so affected will change with the movement of the shoreline.

Alluvium–It is defined as the change in the location of a shoreline by the gradual and imperceptible deposition of soil so as to increase the area of the contiguous land.

Angular Measurement–An angle between two lines at a point is given by the difference in the directions of the lines. The magnitude of an angle can be derived from the division of the circumference of the circle in various ways.

Apparent Solar Time–Apparent solar time is reckoned with respect to the real sun. When the real sun is on the lower branch of the observer's meridian, the apparent solar day is just beginning. When the apparent solar time is 12 hour, the real sun is at upper transit. Hence, local apparent time is the hour angle of the real sun plus 12 hours.

Avulsion–It refers to the sudden and perceptible change in a shoreline, due usually to a river changing its course during a flood or freshet. It is the general rule that avulsion effects no changes in property lines.

Azimuth—The azimuth of a line is the clockwise angle that a line makes with the North end of the selected meridian.

Bearing—The acute angle which a line makes with a reference meridian is called its bearing.

Cartography—It is defined as the science and art of expressing graphically, by means of maps and charts, man's knowledge of the earth's surface and its varied features.

Celestial Equator—The celestial equator is a great circle whose plane is perpendicular to the celestial polar axis.

Celestial Poles— The North and South geographic poles are at the extremities of the polar axis of the earth. The celestial poles are situated at the points where the earth's polar axis extended intersects the celestial sphere.

Celestial Sphere—The stars, which are at enormous distances from the earth, are considered to be fixed on the inner surface of a sphere of infinite radius whose centre is the earth. The observer's position on the earth's surface, or the center of the earth, is considered as the centre of the celestial sphere.

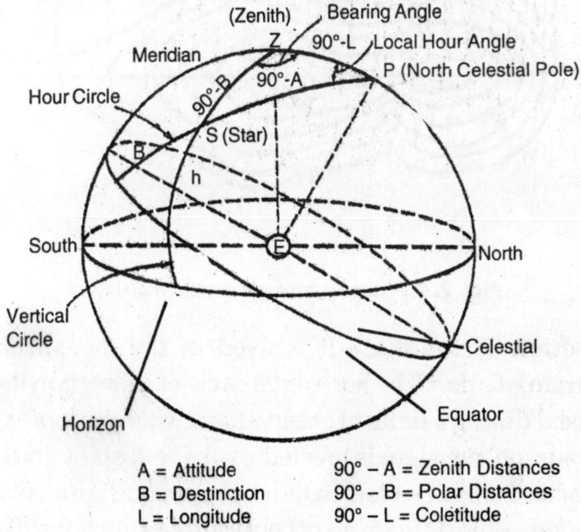

Fig. 1. Celestial Sphere.

Construction Surveying– It deals with the transfer of design dimensions from the engineering drawings to the ground so that the project is built in the correct position and with the proper relationship between its component parts.

Contours– A contour is a line connecting points having the same elevation. It may be considered as the trace of the intersection of a level surface with the ground. The shoreline of a body of still water is an excellent illustration of a contour.

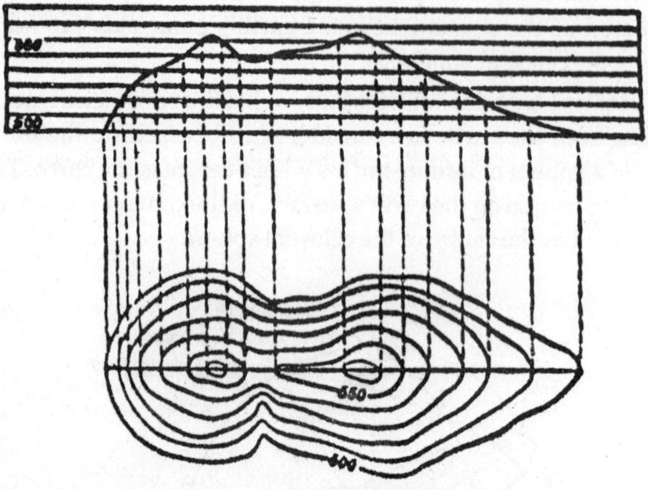

Fig. 2. Contours and ground profile.

Direction Theodolite–It is used in the measurement of angles in triangulation. The horizontal circle of a direction theodolite remains fixed during a series of observations. The telescope is pointed in succession on the signals erected over the distant stations, the direction of each object is read on the circle, and the horizontal angles are calculated as the differences of consecutive directions. The initial direction is arbitrary.

Engineering Surveying—It includes the study and selection of sites for engineering construction, the procurement of design data, and the layout of engineering works.

Fathometer—A fathometer is designed to produce a signal, transmit it downward, receive and amplify the echo, measure the intervening time interval and automatically convert this interval, into feet or fathoms of depth. The fathometer may indicate the depth visually or record it graphically on a roll of specially prepared chart paper.

Geodesy—The science of investigating the form and the dimensions of the earth. It also includes mathematical principle and procedures utilized in the precise determination of the positions of points on the surface of the earth.

Geodimeter— Geodimeter is an Electronic Distance Measuring (EDM) instrument. It uses a laser beam whose invisibility aids pointing on the target. It has an accuracy of ±1 cm.

Global Positioning System (GPS)—Global positioning system consists of a network of 24 satellites which are orbiting in three different planes. Each plane is inclined approximately 63 degree to the equator and it has an orbit period of 12 hours. Each orbit is approximately circular with semi major axes of about 10400 nautical miles. This configuration allows at least six satellites to be simultaneously in view from any point on the globe at any time. Although a single individual satellite passes in and out of view within about 2 hours. Each satellite has its own very stable clock on board and the system is monitored and controlled from a net-work of ground stations. The control parameters include tracking and updating of orbital data as well as data for the timing of the system.

By receiving signals from four satellites, the user determines his positions in three dimension as well as the system time. With GPS it is possible to achieve an accuracy of 5 metres or better in satellite fixes.

Great Circle—A great circle of a sphere is the trace on its surface of the intersection of a plane passing through the centre of the sphere.

Horizon—The horizon is the great circle that is situated halfway between the observer's zenith and nadir points and whose plane is perpendicular to the plumb line. Azimuth is measured in the plane of the horizon.

Hour Circle—An hour circle is the great circle joining the celestial poles and passing through any celestial body. Thus, there is an hour circle for each object in the sky. Of necessity, an hour circle must be perpendicular to the celestial equator.

Jig Transit—It is used principally to establish with precision vertical planes in industrial layout work. It is commonly mounted on a heavy metal stand. It has no horizontal or vertical circles, no compass, and only one horizontal motion.

Latitude— It is the angle between the direction of the plumb line and the plane of the earth's equator. It can also be defined as the angular distance that the observer is North or South of the equator.

Longitude—It is the angular distance that the observer is either East or West, 0° to 180°, from the Greenwich Meridian. Longitude can be expressed either in terms of arc or time measure.

Meander Lines—The traverse of the margin of a permanent natural body of water is termed a meander line. Such lines were run as nearly as possible to conform to the mean high water line. They served the purpose of providing data from which to calculate the areas of tracts of land made fractional by bodies of water. They follow the margins of lakes whose areas are as large as 25 acres, and of streams whose right-angle widths are 3 chains or more.

Mean Solar Time— The mean sun is assumed to move along the celestial equator at the average or mean rate of motion of the real sun. The time provided by the mean sun is such that every day is of exactly the same duration and equal to the average length of an apparent solar day. Hence, local mean time is the hour angle of the mean sun plus 12 hours.

Meridian—The observer's celestial meridian is both an hour circle and a vertical circle, since it passes through the celestial poles as well as the zenith and nadir points.

Fig. 3. Mosaic Construction.

Mosaics—A mosaic is an assembly of photographs of a given area, matched as nearly as possible and pasted upon a background to form a photographic representation of an area.

Nautical Chart— A nautical chart delineates the submarine topography or bathymetry of a given water area by means of depth curves and portrays other significant marine features, as well as those of the shore area. The term nautical chart usually is employed when large and navigable bodies of water area involved. Charts of rivers, harbours, reservoirs and relatively small lakes are frequently known as hydrographic maps.

Orthophotographs—Orthophotograph is a copy of conventional perspective aerial photograph which has been processed to provide the scaling qualities of a map. The displacements of images due to ground relief and tilt of the camera axis have been removed and the photographed features are shown in their true orthographic position.

Photogrammetry—Photogrammetry is the science or art of obtaining reliable measurements of earth surface by means of photography. In terrestrial or ground photogrammetry the

Fig. 4. Photogrammetry, oblique aerial photography.

photographs are taken with the camera mounted on a tripod, and the optical axis of the lens is usually horizontal. Aerial photogrammetry makes use of photographs that have been taken from any airborne vehicle. Such photographs may be either vertical or oblique.

Plane Table Survey— Plane table surveys are operations combining the stadia method and field drafting on a portable drawing board mounted on a tripod. The plane table instrument consisting of an alidade, board and tripod, is essentially a mapping device.

Planimeter— The planimeter is an instrument by means of which the area of a plotted, close figure may be determined directly by tracing the perimeter and reading the result from the scale.

Polaris—It is a fairly bright star (magnitude 2.1) situated about 1° from the North celestial pole. Polaris is a celestial body of great importance to the Civil Engineer and Surveyor.

Profile Levelling—Profile levelling is the operation of determining the elevations of points at regular intervals along a fixed line.

Radian—This is the angle at the centre of a circle subtended by an arc having exactly the same length as the radius. One radian is equal to $360^\circ/2\pi$, or approximately 57.30°.

Reliction—It refers to the gradual and imperceptible recession of a shoreline due either to the rising of the shore or to the subsidence, or drying up, of the body of water.

Riparian Rights—Riparian rights are those vested in an owner by the condition that his property borders on the bank or shore of a stream or body of water. Such rights endow owners with many shore privileges, including the right to construct docks and wharves.

Sextant—The sextant is a portable instrument that is used for measuring the horizontal and vertical angles. In celestial navigation the sextant is used to measure the altitude of various prominent stars above the horizon. In hydrographic surveying the sextant is used to measure the horizontal angles between designated objects on land.

Standard Time— Since the sun does not rise and set in different parts of the country at the same time, it is evident that mean solar time varies from place to place. Standard time is the mean solar time of the central meridian of any country.

State Plane Coordinate System—The basis of this system is the mathematical projection of the earth's surface upon a surface like that of a cone or cylinder, which can then be developed into a plane. There is only one point corresponding to a given pair of plane coordinates, expressed as x and y, for a particular zone of a given state.

Stereocomparagraph—The stereocomparagraph is a stereoscopic plotting device which combines the features of the parallax bar, stereoscope, and tracing arm is one instrument and is capable of measuring spot elevations and of drawing plane metric features and contour lines with greater precision than can be done with the parallax bar apparatus. Thus, the instrument provides a comparatively inexpensive and rapid means of constructing with some accuracy either a plan metric or a topographic map from vertical aerial photographs. The instrument is designed to be attached to a standard drafting arm.

Stereoplotter—Topographic maps are produced with sterioscopic plotting instruments. The optical and mechanical properties of these instrument permit precise control of the floating mark when the stereo model provided by an overlapping pair of aerial photos is viewed and thus relatively accurate contouring as well correct delineation of planimetric detail is achieved. Map compilation constitutes the principal use of stereo plotters.

Surveying—Surveying is the science and art of making such measurements as are necessary to determine the relative position of points above, on, or beneath the surface of the earth, or to establish such points in a specified position. Surveying operations are confined to land, water and extra-terrestrial, space. The essential measurements of surveying are horizontal and vertical distances and directions.

Surveying Monumentation— Monumentation refers to the process of marking in an enduring manner the physical position of survey points. Proper monumentation, together with supporting records and reports, provide the necessary relationship between field surveys and maps and charts.

Surveying Telescope—The surveying telescope is used for fixing the direction of the line of sight and for magnifying the apparent size of object in the field of view.

Tacheometry—The term is generally applied to those operations in which the distance is obtained from one position of the instrument, usually a transit, through the establishment or the measurement of a small angle lying opposite a known base.

Tellurometer—Tellurometer is an electronic distance measuring (EDM) instrument. It provides a fully automatic digital display of distance in metres with an accuracy of ± 5 cm.

Three Arm Protractors—The three-arm protractor is used extensively in hydrographic surveying to plot boat positions during sounding operations. A transparent celluloid protractor consists of a solid disc about 12 inch in diameter containing a circle graduated in degrees, and one fixed and two movable arms extending about 13 inch beyond the edge of the disc. Each of the movable arms has a vernier permitting angles to be set with an accuracy of 2 minutes (2′).

Topographic Map—It portrays by contour lines the shape and elevation of the terrain. In addition to depicting relief, topographic maps show drainage features and the works of man. A variety of symbols and specific colours are utilized to promote clarity of cartographic expression.

Traverse— A traverse consists basically of a series of lines, whose lengths and directions are measured, connecting points whose positions are to be determined. Traverse is the most widely employed operation in the execution of local horizontal control surveys.

Triangulation—This is a method of surveying to determine the horizontal positions of points on the surface of the earth. A triangulation system basically consists of a configuration of triangles all of whose angles have been directly measured.

Trigonometric Leveling—The determination of differences of elevation from observed vertical angles and either horizontal or slope distances is termed trigonometric leveling.

Trilateration—This is a procedure for extending horizontal control which is predicated on the direct measurement of the lengths of all the lines in a geometric configuration and the subsequent computation of the angles.

Tripods—This is essentially a three-legged stand for supporting the instrument and maintaining its stability during observations. It consists of a head to which the legs hinged at the head and pointed metal shoes on the legs.

Vertical Circle—A vertical circle is the great circle passing through the observer's zenith and any celestial object. Obviously it must be perpendicular to the horizon. The altitude of a heavenly body above the horizon is measured along a vertical circle.

Weighted Measurements—The degree of reliability is commonly termed the weight of the measurements. This is merely the relative value of that observation to the others of a series. It is expressed as a number and, being strictly relative, may be multiplied by any factor provided all the others in a series are multiplied by the same quantity.

The general expression for a weighted mean $\overline{M}$ of n observations M_1, M_2, M_3......................... M_n havlbg the respective weights W_1, W_2, W_3,...........W_n is given by the equations

$$\overline{M} = \frac{W_1M_1 + W_2M_2 + W_3M_3 + ... + W_nM_n}{W_1 + W_2 + W_3 + ... + W_n} \quad .... (1)$$

Zenith—The zenith is that point on the celestial sphere where a plumb line extended upward intersects it. The opposite point is termed the nadir.

References

1. Davis, R E., Foote, F.S., and Kelly, J.W., 1966; *Survey Theory and Practice;* McGraw-Hill Book Company, New York, p. 1096 .
2. Dickinson, G.C., 1979; *Maps and air photographs;* Edward Arnold Publishers Ltd. London , p. 348.
3. Greenhood, D., 1965; *Mapping;* The University of Chicago Press, Chicago. p. 289.

4. Kissan, Philif, 1956; *Surveying Instruments and Methods*; Mc-Graw's Hill Book Company. Inc, New York, p. 482.

5. Maling, D.H. 1989; *Measurements from Maps*; Pergamon Press, New York, p. 577.

6. Sandover, J.A., 1961; *Plane Surveying*, Edward Arnold Publishers Ltd; London, p. 424.

7. Schmidt, Milton O., and Raymer, W.H. 1978 Fundamentals of Surveying; D.Van Nostrand Company, New York, p. 450.

8. Schmidt, Milton O., and Wong, Kam W., 1985; *Fundamentals of Surveying*; PSW-KENT Publishing Company, Boston, p. 683.

9. Trorey, Lyle G., 1952; *Handbook of aerial mapping and photograsmmetry*; University Press, Cambridge, p. 180.

Two

Geophysical Field System

Introduction

The function of the geophysical field system is to measure the travel times of seismic waves from the shot point to rock interfaces and back to the geophones in order to calculate the positions of the layers in the earth. It is also to accurately observe the shapes of the echoes in order to drive additional information about those interfaces and their lateral variations.

A geophysical field system receives analog signals at its inputs and records a digital representation of those signals as an output. The number of channels is the number of traces in the spread. The responses of these channels must be identical within the required recording accuracy. This is obtained through careful design and manufacturing but it also requires frequent checking and adjustments on the part of the observer or field geophysicist.

From 1955 onwards, geophysical field systems have undergone a series of modifications in their basic principles. Especially during the last few years, new electronic components have been made available by the manufacturers, enabling improved portability of the equipment, reduction of power requirements and a wide range of possibilities. All these features have led to the fast obsolescence of the earlier geophysical field system. Geophysical recording equipment is a complex electronic system designed by highly specialized electronics engineers.

Geophysical field system is divided into modules regrouping functions. These modules are packed in cases for portable operations or mounted in racks for vehicular operations. Sometimes a choice is offered where the modules can slide either in a truck mounted rack or in their cases. Portability limits the size and the weight of the modules. The heavist one is the tape transport. Advances in technology have made possible to reduce the size and weight of these modules but it should not be at the cost of reduction of resistance to climatic agents and to rough handling. For a given number of channels, the number of modules depends upon the manufacturer and upon the manner in which the functions are grouped. In all cases, the modules whose number varies with the channel capacity, are kept ahead of the multiplexer and the number of modules which function in a multiplexer mode, is kept constant. Besides modularity and portability, another important characteristic of these field systems is the resistance to humidity and water.

Detailed Encyclopaedia

This encyclopaedia is arranged in alphabetical order. The detailed encyclopaedia is given below :–

Accelerometer–A type of seismometer which is responsive particularly to the acceleration of the ground movement. These instruments are generally characterized by high natural periods. Mechanically, they are devices for measuring force or pressure acting on relatively fixed inertia mass.

Acoustic Impedence–A body wave travelling in the first medium and hitting the inter-face splits into a reflected body wave transmitted to the second medium and a wave reflected in the first medium. When the incident ray is perpendicular to the interface, the products ρ V of the density ρ of the medium and the velocity V of the body wave is called the acoustic impedence of the medium.

Active Filter–These filters use active element like amplifier for frequency discrimination. This is achieved by applying selective negative feedback.

Active Geophone–Incorporating electronics in a geophone changes it from e passive transducer into an active transducer. Such a geophone with built-in electronics can put-out a high-level signal into the cable so that the signal is affected less by electromagnetic interference pick-up. The active geophone is powered by the constant voltage between the two output wires, while the current flowing through the wires forms the output signal.

AGC Loop—An AGC loop consists of a rectifier-integrator type circuit fed with the output voltage and controlling the gain of an intermediate stage in the amplifier or a losser element. Some kind of average of the absolute value of the out-put voltage over a sliding time window controls the gain. Adjustments are provided to control the speed at which the system reacts. The compression rate controls the rate at which the gain decreases in the case of an increase of the input. The purpose is to avoid overloading and a faster rate provides more protection. The expansion rate controls the rate at which the gain increases in the decay of the seismic energy.

Air Guns— Air-guns are the most popular type of source for offshore operations. It requires a special valving system to release an amount of compressed air in a very short time. Compressed air is sent to the gun through a hose at a pressure of 140 kg/cm^2(2000 psi) or more. Several factors have an influence upon the frequency content of the pulse created by an air-gun, i.e., water depth, pressure of the compressed air, size of the gun, etc. In general, multiple gun arrays are used.

Amplifier—The functions of seismic amplifier are to amplify the weak seismic signals, to handle the large variation of input signal and to filter out the unwanted signals.

Betsy—The betsy which is a seismic source, is an industrial shot gun. The gun is mounted vertically on a frame carrying a steel plate and a base chamber surrounded by an air-craft type. The base chamber quiets the nozzle blast and minimizes the recoil bounce. The gun shoots 3 ounce (85 gms) slugs at a muzzle velocity of 1750 ft/s (533 m/s), providing an energy of 9000 ft-lb (12.2 kJ).

Binary Gain Amplifier—The gain of a binary gain amplifier changes in 6 db steps. The output of each amplifier is continuously compared with two reference voltages an upper and lower signal threshold. When the amplifier output level exceeds the upper threshold a gain decrease command is generated and when it falls below the lower threshold, a gain increase command is generated. 90 db bidirectional gain ranging system of a binary amplified is capable of recording the exact gain of the amplified at any time without distorting the character of the reflected signal. The ranging ability for an up-step usually occurs once every 30 sampling intervals. The ability to step-down is commonly available at one step or 6 db after each sampling interval.

Blaster – A blaster is essentially a source of current connected to the firing line. It is expected to fire all the blasting caps of the circuit within one millisecond of each other or so. The voltage capability of a blaster is 200V or more. The energy required for firing the caps is stored in capacitors and released at the shot instant. The capacitors are charged by a battery or a magnet. Large capacitors are used to store a maximum of energy. A voice communication system is incorporated in the blaster unit. It can be a sound powered handset through which the shooter communicates with the observer or an intercom amplifier. The blaster receives a fire signal from the recording instrument and returns a time break signal along the same transmission line. In the case of shot holes, up-hole has to be recorded. An output is provided to connect the up-hole geophone to the blaster and the signal is transmitted through the communication system.

Booster – A booster is a small load of a high class explosive. It is detonated by the blasting cap and its detonation develops much more energy than the cap alone, enough to detonate the main mass of low value explosive. Explosive cords, such as primacord, can be used as boosters.

Cable Limitations – In conventional seismic systems, each geophone is connected to the recording system by one pair of conductors and all pairs are bundled into a multi-pair cable. These cables are sectionalized for ease of transportation and each string is connected to the next by plugs which must carry as many contacts as there are conductors in the cable. To limit the weight and size of these plugs, each contact has to be made as small as possible. When the number of pairs increases then the plug weight as well as linear weight of the cable increases. Repair time also increase with the increase in the number of pairs in the cable. High initial cost and high maintenance and repair cost set a practical limit to the number of fairs in a cable. The transmission of analog data along cables is a source of crossfeed, of distortion, and of noise pickup.

Common Mode Rejection Ratio (CMRR)—The ratio between the output voltage generated by a differential input and the output voltage generated by the same input voltage applied as a common mode to both inputs is called the common mode rejection ratio.

Components Inserted P.C. Board—The components are plugged into sockets which are soldered on the board. An efficient soldering is achieved by automated means such as wave soldering in which the tin alloy creeps up the pins into the plated-through-holes and provides a maximum cohesion. The pluggable elements are available in the form of integrated circuits. The module is made of a small PC board on which the active and passive elements are permanently soldered. It is encapsulated in a waterproof box, filled with epoxy, and its terminals are pins perpendicular to one face which plug on the main board. Sometimes the pluggable element is even bigger and constitutes an auxiliary board attached to the mother board and connected to it by some type of terminal plug.

Correlator—In vibroseis, operations, the reflections are not apparent on an uncorrelated record and the camera alone cannot fulfill its monitoring purpose. A monitor cross-correlated record on which reflections are visible can be obtained by the use of correlation in the field if only a few bits of the data and/or pilot signal are retained for the operation. Most correlators retain the sign and the 7 most significant bits in the sweep and in the data signals for cross correlation. Each correlator channel contains a memory, a multiplier, an adder and accumulator. The sweep is stored in a separate memory of appropriate capacity. At the end of the first scan, the accumulator contains one sample of the cross-correlation function. At the end of the successive scans, the corresponding samples of the cross correlation functions are available in the accumulators. These samples are converted into analog signals and the filtered outputs are applied to the camera for display in real time without losing the production time.

Correlator Stacker – It is possible to perform a full precession cross-correlation in real time, before vertical stack. The correlator-stacker performs all the operations of the stacker and of the correlator, which are described separately. The output of the correlator-stacker is digital and applied to the formatter to be recorded on tape. Monitor displays (records) are available through the read-after-write and playback facilities of the recording system.

Counters – A counter is a group of flip-flops in which the change of state of any one occurs when all previous flip-flops are in a set condition.

Cross-feed – Cross-feed is generated by the coupling between channels through leakage and through induction of currents circulating in one circuit and creating an electromotive force in others. In this manner, the noise and signal of one channel appear more or less attenuated in other channels and mix with the signals of these other channels, constituting a noise. Crossfeed is measured by the ratio of the signal in the channel leaking to the other to the signal induced in the second channel. The non-excited channels must be terminated on a specified resistance at their inputs. Crossfeed is usually expressed in db.

Data Block – The data block is made of successive scans, each scan corresponding to a cycle of the multiplier, to a sampling interval and auxiliary channels during that interval and containing the samples taken in all channels, seismic and auxiliary channels, during that interval. Each scan starts with a synchronization group. The first start of scan is coincident with the time break. The start of scan indication is unique and constitutes a group of characters which cannot be found anywhere else in the data values. The data block is written according to the selected format. All seismic formats include a special start of scan and express the data in the binary system.

Digital AGC – In a floating point system, the deformated data consists of a mantissa and of a gain exponent. The mantissa alone cannot be used for a monitor display and a gain recovery is necessary to bring all the samples of a track back to the same scale, to the same gain, before assembling them into a continuous analog signal. The gain recovery is accomplished by a 28 bits shift register. After the gain recovery, the variations in the data values are 80 db or more for any given seismic trace. The dynamic range of an oscillographic camera is only 20 db. Therefore these variation in large data values require a dynamic compression through an automatic gain control. In this case AGC is applied to sequences of digital samples, therefore it is called a digital AGC.

Digital Field Systems – Seismic data acquisition system is divided into modules regrouping functions. Those modules are packed in cases for portable operations or mounted in racks for vehicular operations. Sometime a choice is offered where the modules can slide either in a truck mounted rack or in their cases. Potability limits the size and weight of the modules. The heaviest one is the tape transport. Advances in technology have made possible to reduce the size and the weight of these modules but it should not be at the cost of a reduction of resistance to climatic agents and to rough handling. For a given number of channels, the number of modules depends upon the manufacturer and upon the manner in which the functions are grouped. In all cases, the modules whose number varies with the channel capacity are kept ahead of the multipler and the number of modules which function in a multiplexed mode is kept constant. Besides modularity and portability, another important characteristic of these field system is the resistance to humidity and water.

Digital Seismic System – The digital seismic system receives analog signals at its inputs and records a digital representation of these signals as an output. The number of channels is the number of traces in the spread. The responses of these channels must be identieal within the required recording accuracy. This is obtained through careful design and manufacturing but it also requires frequent checking and adjustments on the part of the observer.

Digital Tape Formats – Many arrangements can be imagined by which to record seismic data on tape. Digital Recording Standards Committee set up by SEG has standardized only 9 track formats, though acknowledging the existence of 7 and 21 track formats. The recommended formats are named by the letters of the alphabet, the first letter being used for the multiplexed formats. Different type cf formats are:- SEG-A, SEG-B, SEG-C, SEG-D and SEG-Y. Popular format SEG-B is applicable to data recorded with binary gain or IFP amplifiers. The gain exponent is recorded in full and complete data words are packed in 2.5 bytes.

Dinoseis—Dinoseis is a gas exploder and it is used for as a seismic source. It contains a wide and heavy piston which, with the steel pan constituting the bottom of the assembly, encompasses a combustion chamber. This chamber is filled with propane and oxygen from bottles carried by the truck and the mixture is fired by a regular spark plug. The flexible steel pan transmits the energy to the ground. The free piston is thrown upwards, compressing an air cushion between itself and the upper part of the bell. In order to provide enough energy for deep seismic surveys, several gas exploders are used together and they are popped several times.

Dynamic Range—The dynamic range of an amplifier is given by the ratio of the maximum and minimum linear inputs which it can handle within the specified distortion in the output.

Explosives—An explosive is a chemical compound or mixture capable of liberating its energy in a very short time. In general, the reaction liberates a large volume of hot gases which exert a strong pressure on the surrounding medium. Most often, the reaction is a combustion involving oxygen bearing components and easily oxidized elements such as carbon and hydrogen. Different type of explosives are: geletin dynamites, Ammonium Nitrate Fuel oil, nitroglycerine, etc.

Filters—A filter is a circuit which transmits a range of frequencies unaltered and does not transmit the others. Ideally, the transfer function has a finite constant module at the frequencies allowed to pass on and a zero module at the others, presenting a sharp step at the limit of the pass-band. The phase characteristic is linear and all frequency components are delayed by the same amount of time which is the filter delay. But in reality the amplitude response at the allowed frequencies is rnade as nearly constant as possible and smaller than a given attenuation figure at the rejected frequencies.

Flip-Flop—A flip-flop is a circuit whose output sitches from one state to other when it receives an impulse. Flip-flops are used for storing digital information. They store a bit one when they are set and a bit zero when they are in the reset state.

Formatter—The digitized data coming from the A/D converter and the gain information coming from the gain ranging amplifier are forwarded to the formatter where they are set in a form complying with the recording format requirements. The formatter is made of logic circuits in which the out-put of the A/D converter is made ready for recording operations. A seismic format requires each word to be divided into several characters. The incoming word is separated into smaller group of bits and each group is stored in a separate register from where the data can be retrieved when necessary. Also in the formatter special characters are generated, e.g., start of scan words, the time words, end of file marks, etc.

Gain—When the input resistance and output resistance of an amplifier are equal then the gain of an amplifier is equal to the twenty time the log of voltage amplification in db. This is given by the expression:—

$$G = 20 \log_{10} A \text{ db} \qquad (1)$$

Where A is the voltage amplifteation of the amplifier.

Geometrical Spreading—As the seismic wave travels through homogeneous medium, wave fronts move in spheres whose surfaces vary as the square of the radius. So per unit surface, the energy decreases as the square of the distance. Amplitudes are inversely proportional to the distance travelled. This is called geometrical spreading. In geometrical spreading the total energy remains unchanged.

Geophone Array—The signal to noise ratio is improved by replacing a single geophone by an array of several geophones scattered over a length or over an area around the central trace location. Linear arrays provide a filtering effect in the space domain.

They enhance the vertically emerging reflected energy more than the horizontally travelling noises. Multiple geophones also provide a statistical improvement of the signal to noise ratio by reinforcing the systematic signals more than the random noises.

Geophone Distortion—The distortion is that which occurs when a geophone fails to reproduce precisely the signal which excites it. Thus an undesired change in the output waveform is called distortion.

Geophone Ground Coupling—It refers to the accuracy with which thhe geophones measures the actual ground motion. The large amount of information carried by the seismic signal can only be brought into play for seismic interpretation if geophones are able to follow the movement of the ground faithfully. The geophone response is defined as the output voltage of the geophone, expressed as a function of frequency, in response to ground motion for which the velocity is a constant function of frequency.

Geophone Grouping—Geophones are deployed in groups at each detector location. The group is called the string or an array. Each string is attached to a take-out in the main cable which feeds to a specific channel in the seismic recording system. In a multielement geophone string, the geophones may be connected in a series, parallel, or in series-parallel combinations.

Geophone Noise—There are two major sources of noise in a geophone. One is the thermal noise due to coil resistance. The other is due to the Brownian partical motion in the gas within the geophone case.

Geophone String—The number of geophones per trace varies commonly between six and a few tens. It would not be convenient to connect each geophone separately. So, groups of six to twelve geophones are permanently connected in strings along a leader cable and the array includes a variable number of strings connected to a centre point or along one or several inter-connector cables. The number of geophones in a string and their mode of connections must allow a variety in the number of strings connected together at a geophone station without impairing the response of the group, without deteriorating the damping factor.

Harmonic Distortion—Harmonic distortion is characterized by the appearance, at the output of a seismic system, of frequencies which are integral multiples of the input frequency. The input frequency present in the output is called fundamental frequency and other frequencies present are called harmonics. Harmonic distortion is generated by the non-linearity in the recording system. The total harmonic distortion is the square root of the ratio of energies carried in all harmonics together with the energy carried by the fundamental frequency. It is expressed in per cent. Its order of magnitude is 0.05 to 0.1 per cent.

Header Block—The header block contains the identification pertaining to the seismic record. It is a short record including standard information which is supposed to follow a recommended pattern and an optional part in which each user may add other data at will provided the system is equipped with the means to introduce this additional information on the tape. The identification of the record and the parameters used for recording are all coded in BCD for storage on tape and occupy four bits per decimal digits. Two digits are associated with each character, making an eight bit byte to which an odd parity bit is added.

Hybrid Circuits—Hybrid circuits are a combination of conductive and resistive elements in the form of deposits made on a surface and of independent components bonded to this base. They offer the possibility of miniaturization of the analog circuits. The substrate is a ceramic material. Metallic conductive connections are laid on the substrate using masks to delimit their surfaces.

Hydrophone—A hydrophone is a pressure sensitive device. It converts the pressure wave generated in water into a voltage signal. The output of the hydrophone is proportional to the acceleration of the generated waves. Hydrophone consists of a piezoelectric element that converts changes in water pressure into electrical impulses, which are transmitted to the recording unit of the marine survey ship.

IFP Amplifier—In instantaneous floating point amplifier the gain ranging system consists of attenuators and FET switches to perform the 12 db gain steps. Logic levels are used to control the gain switches. Each step of gain ranging system takes approximately 3μs. The attack time and release time of an IFP amplifier are 4×10^6 db/sec. The basic objectie of the IFP amplifier is to adjust the amplifier gain for each seismic data sample to the highest value possible without overloading the A/D converter or overloading any amplifier gain stage.

Input Noise—In an electronic system, an output can be observed even in the absence of input. This is called system noise. In a seismic data acquisition system there are noise generators and amplifier stages at several locations. Depending upon their relative positions the noise recorded at the output presents different values. But, given a total system gain, this output could be obtained with a perfect system without input signal. This hypothetical input is the input noise. To express noise, the voltage observed at the output in the absence of an input is divided by the overall gain so as to obtain a voltage referred to the input.

Integrated Circuits—The gates, flip-flops and other elementary circuits resulting from their combinations are available as integrated circuits in the shape of small parallel epipedic packages or encapsulated in cylindrical boxes. The circuits are laid on a piece of crystal of silicon. Semiconductor junctions are by created introducing impurities into the silicon. Resistors and conductors are obtained by metallic deposits. Other deposits provide the insulation between conductive layers. The area over which each deposit is made is limited by a mask. This small circuit constitutes a chip of surface area of 1.25×1.25 mm^2 and is encapsulated in metallic package. The ports of the circuit are connected to accessible external terminals. The most common package is DIL (dual in line) package in which the terminals are distributed along two lines on the sides of a parallelopiped of epoxy containing the chip.

Jack Kilby of Texas Instruments in Dallas had shared the Nobel Prize in Physics for the year 2000 with Zhores I Alferov, Russia and Herbert Kroemer, USA. He got half of this prize for the invention and development of the integrated circuit, the forerunner of the microchip and as a co-inventor of the pocket calculator.

The other half had been shared between Zhores I Alferov and Herbert Kroemer for their work in developing technology used in satellite communications and Cellular Phones.

Laser Camera—The light beam produced in a helium-neon laser is projected over a rotating mirror and scans the width of the paper. The bear is modulated by a microprocessor controlled electronic system. As the mirror sweeps the beam across the paper, this beam is turned off and on as required to produce a raster of 0.005 inch dots, each appropriately positioned to form a part of the image to be displayed. The complete image is formed by a continuous series of rasters at 0.005 inch intervals along the record. The paper is driven by a stepper motor synchronous with the raster at 0.005 inch per scan. The input of the laser camera is derived from the output of the digital AGC in the playback circuits of the recording instrument. In a laser camera a high number of channels can be displayed.

Longitudinal Wave—The longitudinal seismic waves are called P-waves. In longitudinal waves the direction of particle motion is along the direction of wave propagation. The P-waves are sound waves and they are faster and contain higher frequencies. These waves can be generated easily and these waves can travel in both solid and liquid.

Magnetic Heads— The bits of a character are recorded by a bank of magnetic heads over which the tape is driven. A magnetic head can he considered as an electromagnet. The direction of the magnetic field depends upon the direction of the current in the coil. At the gap, the magnetic fiux selects a path of lower reluctance across the oxide instead of across the air in the gap. A portion of the tape is magnetised. Each oxide granule becomes a small magnet. When that magnetized tape is driven again over a magnetic head, the variation of the fux induces a voltage in the coil.

Magnetic Tape—The seismic data is recorded on a magnetic tape half an inch wide. Seismic data are recorded as a remanets magnetic induction in iron oxide. The oxide is made of tiny elongated granules held with a resin glue on a mylar support. The thickness of the oxide coating is 1/2 mil and the total thickness of the magic tape is 1.5 mil. Computer tapes come in 2400 ft. lengths on $10\frac{1}{2}$ (10.5) inch diameter reels. To increase their portability, many seismic field recorders are made for smaller 8 inch reels with a capacity of 1200 feet of tape. All these reels have the same hub and can be mounted on any tape transport, in the field or in a computer centre.

Microprocessor Controlled System—The operations carried out in seismic data acquisition system include a number of logic functions which have to be performed in due time, as a well established sequence. The sequence could be controlled by a computer instead of being hardwired. Software controlled offers more possibilities than hardwired logic but, because of the flexibility, it is slower. Microprocessors offer the flexibility of software control. Besides, they are small and easy to incorporate in a seismic system. The processor performs instructions one at a time. The programs are stored in a memory the content of which can be modified. Microprocessors with 16 bit words or 32 bit words are suitable for seismic recording. In microprocessor control seismic systems, the results of the periodic test are immediately available to the observer so that he can decide accordingly. Also, different sequences of data acquisition can be programmed and called when necessary, allowing a wide flexibility in the operation of the system.

Mini Sosie—Mini-Sosie is a low cost go anywhere, shallow exploration tool. The Mini-sosie makes use of a sequence of discontinuous impulses of long duration. It is assumed that all pulses from all sources used simultaneously are identical. If this sosie response is cross-correlated with the sequence of pulse instants, the result reduces to the conventional seismic trace. The pulses are produced by one or several gasoline powered earth tampers. By constantly moving the throttle, the operator varies the rhythm of the engine and obtains random impacts upon the ground surface. A sensor attached to a tampering plate translate each impact into a pulse which is fed into the Mini-sosie system by a wire or a radio connection.

Mini-sosie Processor—A Mini-sosie processor can be viewed as an adder and a memory controlled by a microprocessor. It is connected to the amplifier outputs. Each new sample coming out of the amplifier system is added to the content of a memory address controlled as a function of the time of each new pulse of the earth tamer used as a seismic source. The Mini-sosie processor operates with 12 bit words.

Multi-layer Card— A multi-layer card is made of several thin sheets of dielectric material on which circuits are printed. These sheets are separated by insulating layers and assembled into a single board by a polymerization process under controlled temperatur and pressure. Connections between the different layers are established by plated-through-holes. Repair of multi-layer card is very difficult.

Multiplexer—A multiplexer is a kind of selector switch connecting several inputs successively to a single output. The sliding contact, turning at a constant speed, connects inputs 1, 2, 3, etc one after the other and in this order to the output. After the last input, n, input No. l is connected again to the output and the sequence repeats. A multiplexer is made with FETS electrical switches on the inputs with a logic control to switch one of them ON at a time.

NRZ Recording—The binary number system is selected for data expression because it fits the alternative possibilities or electronic system very well. Bits are coded according to Non-Return to Zero (NRZ). Codes which require only two magnetic states which are positive and negative. Write current is in one direction for bits 1 and in other for bits 0. A change occurs when the bit changes and the code is called NRZ-C. There cording code mostly used in seismic survey is NRZ-1. In this case, the write current is inverted each time a bit 1 occurs. When this tape is read, pulses occur each time a bit 1 was recorded. The decoding is very simple. In the case of a pulse, a bit 1 is generated and, in the absence of pulse, a bit 0 is generated.

Offset—The offset is generated mainly in the switching organs of the system, in the multiplexer and in the floating point amplifier. Its effect is the reduction of the maximum input voltage. If this voltage is V_1 in absence of offset, the amplitude of the largest acceptable signal becomes $V_1 - V_0$ when the offset is $+V_0$. When V_0 is positive, signal amplitudes larger than $V_1 - V_0$ would produce an output voltage bigger than V_1 at the peaks and cause saturation or excessive distortion. When V_0 is negative, saturation would occur in the troughs. The dc offset shifts the zero amplitude axis but the upper and lower voltage limits stay in place. As seismic signals have a zero average, they can be easily separated from the dc offset by computing the average of all samples of a seismic trace and by subtracting this average from each sample.

Operational Amplifier—Operational amplifiers are used in anolog circuits. They are called so because they were developed to perform mathematical operations in analog computers. They are based to the principles of negative feedback. The open loop again of the feedback amplifier is very high. Ideally the input impedance of the feedback amplifier is infinite so that it does not constitute a load for the input circuit and the output impedance is zero so that the output voltage is not affected by the output load. Operational amplifiers are widely used for various functions in the seismic systems, e.g., differential input amplifiers, filters, comparators, amplifier stages, etc.

Orthogonal Recording—Formally the displacement of the earth at any point due to propagation of seismic waves have three dimensional components which are measured by three orthogonal geophones consisting of one vertical and two horizontal geophones. Their operational axes are perpendicular to each other. These vertical and horizontal geophones are mounted on a solid cast block and are sealed in an aluminium case.

Oscillographic Camera—The function of the camera is to monitor the field record quality in order to ascertain, if reflections are obtained from the required depth. The voltages representing the seismic signals are applied to moving coil galvanometers which carry mirrors. The galvanometers are inserted in a permanent magnet, and the assembly constitute the galvanometer block. The mirrors deflect a light beat which is focused on a strip of photosensitive paper driven at a constant speed. The deflection of light beam is proportional to the input voltage and, on the paper, the deviation of the light spot represents the seismic signal. Each galvanometer generates one trace on the paper. Timing lines are generated by a timing lamp at regular time intervals.

Parasites – These are unwanted signals generated within one geophone due to horizontal and tortional motions. These depend upon the construction of the geophone.

Pareto Principle – In quality control, the bulk of the field failure and downtime are traceable to a vital few field failure and defects. This phenomenon of vital few and trivial many is universal and is known as the Pareto Principle. A major use of Pareto Principle is in the design of quality improvement programmes. Improvements can be justified only for the vital few defects and these are identified by listing of all the defects in the order of their magnitude and occurrence.

Passive Filter – An electrical filter is frequency selective network. Passive filters use resistive, inductive and capacitive elements for frequency discrimination.

Passive Geophone – Most exploration geophones are passive electromagnetic transducers. They consist of an electrical part, comprising one or two coils and a magnetic part, comprising a permanent magnet and a magnetic circuit with one or more gaps. The coils are placed in the gaps and the two parts can move with respect to each other.

Phase Encoded (PE) Recording – Phase encoding recording provides a means to reconstruct the write clock, even if the tape speed is different at the write and at the read times. In the split phase encoding system a flux inversion occurs in one direction for a bit 1 and in the other for a bit 0. As a consequence, another reversal is necessary when two similar bits follow each other. If we consider the frequency of occurrence of the flux reversals, the record along one track is made up of two frequencies only, one being twice the other when identical bits follow each other in a sequence. The lower frequency is that of the write clock. The split phase code is used for 1600 bpi PE records. In the bi-phase mark encoding system, a flux reversal occurs at each bit with an additional reversal when the bit to be represented is 1. There again, the record is made up of two frequencies, one being twice the other where the bits are 1. The lower frequency is that of the write clock. This type of encoding is used for high density digital recording at densities of 8000 bpi.

Poissons Ratio—It measures the compressibility of the medium and it is influenced by variations of the fluid in the formation. It varies when the matrix contains gas instead of water and the variation of the reflection coefficient is responsible for the appearance of a "bright spot".

Preamplifier Filters—The front end of a seismic data acquisition system consists of a number of preamplifier filters equal to the number of seismic channels. High level low frequency noise such as ground roll must be attenuated before reaching the automatic gain ranging system. Otherwise, valuable signals weaker than the noise could be compressed below the instrument noise to a non-recoverable level. The gain ranging amplifier is located beyond the multiplexer. If filtering is to be performed on seismic signals, the low cut filter must be located ahead of the multiplexer, in the preamplifier. The seismic signals are sampled, therefore high frequencies which could generate aliasing have to be eliminated by a high-cut filter. The aliasing filter has to be located in the preamplifier. High voltage transmission lines generate noise at line frequency (50 or 60 Hz) in the seismic band; and this noise has to be attenuated before gain ranging. For this reason, a notch filter is provided in the preamplifier.

Quality Control— Quality control is the regulatory process through which we measure actual performances of the geophysical instruments; compare them with the design specification of the manufacturer's and act on the differences. The goal of any quality control programme is that of quality assurance and prevention of defects.

Printed Circuit Boards—Active as well as passive elements are placed on boards and their terminals are connected by conductive lines printed on the board. A board is made up of a dielectric material on which conductive surfaces or lines are deposited. These surfaces or lines are made of copper and covered by a tin-lead alloy which makes it easier to solder components on the board and protects the copper deposit against corrosion by chemical agents. Both faces of the board carry printed interconnections. Connections have to be established between the conductors on both faces. They consist of plated-through-holes, and coated with copper and tin-lead alloy. These holes are used also for fixing the components on the board.

Radiation Impedance—When a geophone resting on the ground surface begins to vibrate in response to an incident compressional wave, the ground will exert a reaction force on the geophone. The complex ratio of this reaction force to the velocity of the geophone is defined as the geophone self mechanical radiation impedance.

It consists of a real part known as a radiation resistance which is a measure of the re-radiated real power of the incident compressional wave.

Radio Firing—CDP techniques often require shooting from several locations with relatively short time intervals to avoid losing time while the recorder stays in one place. Repeated moves of the shooter are made easier by radio link which must provide voice communication, the transmission of the fire signal from the recording end to the shooter and the transmission of the TB (time break) and up-hole geophone signal back to the recorder. The transceiver is coupled to the seismic source controller at the source end. At the recording end, the interface generates a signal when it senses a fire order received from the seismic data acquisition system. The signal transmitted is in a code form. The code is detected by a decoder at the source end and it triggers the blaster. When the source is initiated, the time break and up-hole signals are modulated in the decoder module to be transmitted to the recorder over the radio. At the recorder end, the encoder module demodulates these signals and applies them to the appropriate seismic system inputs.

Recorder—In seismic recorder data and gain bits coming from formatter are translated into currents which generate magnetic flux inversions on the magnetic tape. All seismic recording are made on nine tracks by nine magnetic heads.

There are a write amplifiers which write 9 bits simultaneously on the tape. In tape deck the tape is driven over the write heads by the mechanical system. The seismic recorder is also used to replay the tape. This tape generates emf at the terminals of the 9 read heads and these emf are converted into bits by 9 read amplifiers. The tape driven system generally includes one motor for each real and one motor for capstan driving the tape at controlled speed in the vicinity of the head bank. Controls are provided on the tape deck to move to the tape in one direction or the other.

Recorder Limitations—Recorder set a severe limitation to the number of traces which can be recorded in seismic surveys. One way to escape this limitations is to use a higher density, non-standard, recording code, as can be done with a Gas-Bus system, and to provide a special tape reader at the input to the processing centre. Another way is to reduce the number of bits per word, as is done in the Geocor system, in which case the advantages of recording more channels may compensate for the loss of accuracy. Better way is to avoid recording in real time, to buffer the data in one or several intermediate memories, solid state or cassettes, and to retrieve it later on at a reduced data rate, if this additional delay is acceptable.

Registers—Bytes or words are stoked in resistors. They are bank of flip-flops. The number of flip-flops in the assembly determines the capacity of the register.

Reliability—Reliability is a property of geophysical instrument to fulfill all its given functions under certain conditions of operation during a given time while retaining its basic parameters within the limits stated in specifications.

RMS Voltage—The rms voltage of a periodic or non-periodic signal is the dc voltage, which applied to the same resistance, would dissipate the same amount of energy. The rms amplitude of a sine wave is the peak amplitude divided by $\sqrt{2}$.

Roll Along Switch—CDP stacking requires shooting at short intervals along the line while moving the spread along the same reduced distance, discarding one or a few traces at the back and including the same number of additional traces at the front end. To avoid loosing time, more traces than necessary are laid in the field and connected to the recording unit where only those which are to be recorded are connected to the system inputs. The interface between the spread cables and the instrument inputs is the roll along switch. The number of outputs is the number of recording channel of the data acquisition system. They are carried by elements providing moving contacts with the inputs. As these elements are moved, the correspondence between input and output channels is altered and the spread actually connected to the recording instrument is moved along. Roll-along switch is made of sliding contacts distributed along a circle. The sectors carrying the outputs to the recorder are rotated either manually or automatically, at the receipt of a signal from the recording unit. Many roll-along switches provide a positive indication of the position of moving blocks and this indication may be coded in the header record.

Sample and Hold– The samples or elements of signals at the output of the multiplexer follow the variations of the input signals. But the digital conversion is not instantaneous. It involves a comparison and cannot reach or meet the required accuracy if it is not carried on a constant signal. Therefore, the output voltage is kept constant during the sampling time by sample and hold circuit. This output is stored on a capacitor which has to be connected to the multiplexer output at one time and to A/D converter at another time. We can schematize the sample-and-hold as a capacitor between two gates. As there is little time to load the capacitor to the new value, the time constant of the circuit including the capacitor and the input must be short. On the contrary, to maintain a nearly constant charge at the output, the time constant of the capacitor and output circuit must be very large. To meet these conditions, the input impedance must be extremely small and the output impedance extremely large. The first sample and hold is located between the multiplexer and gain ranging amplifier. A second sample and hold is located between the gain ranging amplifier and the A/D converter.

Seismic Cable– The seismic cable is a multi-conductor cable with one pair per trace. Seismic cables carry the signals from the geophone groups to amplifiers. Each cable section is provided with one or several takeouts regularly spaced at intervals slightly longer than the longest trace interval considered. A takeout is the element through which a group of geophones can be connected to a pair of wires in the seismic cable.

Seismic Instruments—The function of the seismic instruments is to measure the travel times of seismic waves from the short point to rock interfaces and back to the geophones in order to calculate the positions of the layers in the earth. It is also to accurately observe the shapes of the echoes in order to derive additional information about those interfaces and their lateral variations.

Seismic Record—A seismic record is the data obtained from one shot or a similar type of energy release, picked up by a spread of geophones. A seismic record is made up of the representation of the geophone voltages called data and of another part made up of recording parameters is called a header. A seismic record constitutes a file on the tape. Instead of intermixing the words from different channels, the data from each channel are separated and recorded in different blocks. In that case, there is one general header block followed by files of seismic records. Each zone of these last files is made of one record per trace with the trace identification as a first part of this record.

Seismic Source—A mechanical disturbance is generated at the origin of a seismic observation by seismic source by displacing momentarily a small volume of rock from its rest position. Seismic energy can be generated by dynamite, solid chemicals, compressed air, gas exploder, implosion, electrical method, weight drop, vibrator, etc.

Seismic Waves—Seismic waves are elastic deformations which propagate through the earth. The initial deformation is that created at time zero of the seismic record when the source moves a small volume of rock out of its rest position by exerting a pressure on it. In the case of vibroseis, the deformation is continuously controlled over its finite duration. In the case of impulsive sources, the pressure of the shock wave is released after its application and, beyond a small region of permanent deformation, the particles disturbed by the pressure wave return freely to their initial positions under the effect of elasticity.

Shear Wave—The transverse seismic waves are called S-waves or Shear Waves. In shear waves the direction of particle motion is perpendicular to the direction of the wave propagative. Shear wave velocity is about half of longitudinal wave velocity. These waves can travel only in solids.

Shift Registers—A shift register is a register in which the data can be moved in one direction or the other. On receipt of a clock pulse, a flip-flop of the register takes the state that its left or right neighbour had before. Shift registers perform multiplication or divisions by a power of two.

Sign-bit Recording—Sign-bit data recording is a technique for acquiring seismic data by converting the analong geophone signal to a one bit digital representation. The one bit representation reflects the polarity or sign of the analog geophone signals at each sample time. The relative amplitude of seismic events is regained from sign-bit data by cross-correlation with a reference signal or by vertical or horizontal summing when typical seismic noise is present.

Slew Rate—Slew rate is defined as the time rate of change of output voltage of a unity gain amplifier for a step input. This is given by the expression

$$\text{Slew Rate} = \frac{\Delta V}{\Delta t} \text{ volt}/\mu \text{ sec.} \qquad ....(2)$$

where V is the change of output voltage at Δt time.

Sparker—The main application of the sparker is for shallow underwater surveys and it is used with electrochemical recorders synchronized with the triggering of the spark. A spark discharge between two electrodes produces a load, short duration sound. In the water, steam is generated by the energy and expands in a bubble creating a seismic wavefront. The energy is accumulated in capacitors prior to the shot. Voltages of a few thousands to a few tens of thousand volts are applied to electrodes towed behind the boat. The peak acoustic energy is produces between 250 and 600 Hz and significant energy is present up to 5000 Hz.

Stacker—Vertical stacking consists in adding corresponding data words obtained in successive data acquisition cycles without moving the spread. The stacker consists of a summing device and a memory in which the vertical stack can be made. The stacker is interfaced in the data acquisition system between the A/D converter and the formatter. The seismic source is activated a number of times corresponding to the vertical stacking fold and, each time the memory content is updated. When the required number of cycles have been accomplished, the memory contains the full stack which can be dumped on tape to generate the stacked record at processing center. A stacker has to be interfaced with the data acquisition system. The interface card is exchanged when the stacker is to be used with another type of data acquisition system.

Streamer— Streamer is used for offshore operation. It includes the hydrophones, the signal transmission cables and the stress number which withstands the traction strain. It is towed behind the survey ship at a constant speed, the seismic source is actuated at constant intervals and recording is performed as the survey ship and streamer move along. It is contained in a hose making a cylindrical jacket 2 to 2.5 inches in diameter. The streamer is filled with a low density fluid so as to obtain a neutral buoyancy.

Surface Waves (Ground Roll)—They propagate radically along the surface, around the shot point, at a velocity much lower than that of body waves, usually less than 1000 m/s. Another characteristic of surface waves is their low frequency content, mostly below 20 Hz. Because their frequency spectrum is different from that of the reflected energy, part of them at least can be attenuated by a low cut filter. To prevent the large amplitudes of surface waves from controlling the gain, the low-cut filter must be located ahead of the gain ranging device.

Synthetic Seismogram—The outer part of earth, especially in sedimentary basins, is made of numerous layers, each presenting special elastic properties, which can be characterized by their acoustic impedances. The sequence of acoustic impedances along a vertical is the log where ρ is the density of the medium and v is the velocity of the seismic wave. From the ρv log, it is possible to compute a log of the reflection coefficients. If we consider a vertical ray-path, this log, convolved by the initial disturbance, provides a synthetic seismogram without multiples and without the effects of attenuation. These effects, as well as multiple reflections, can be taken into account for the computation of the synthetic seismogram so as to arrive at a computed reflection signal which can be compared to the recorded signal.

Tape Configuration—On a tape data blocks are written. A file of information can be made of several records or data blocks. Data blocks are separated by inter record gap. Reflective markers are located along one edge of the tape at not less than 10 feet from each end. They constitutes the beginning of tape (BOT) and end of tape (EOT) markers. Those 10 feet of tape at each end constitute trailing ends. They are not used for recording and they can be manipulated by operators to load or unload the tape. Tape transports can read either 800 bpi NRZ (Non-return to Zero) or 1600 bpi PE (phase encoding) or 6250 bpi GCR (Group Coded Recording). They switch automatically into the appropriate read mode, after sensing an identification burst at the BOT. Files on a tape are separated by end of file marks or tape marks. When the file is the last on a reel, the tape mark can be repeated to indicated that this is the end of the recorded part of the tape.

Tape Skew—The bits of a character are recorded with different magnetic heads, one per tracks, assembled together into a head-bank. Manufacturers attempt to align all gaps on a straight line but the alignment is physical, and it is not perfect. There are discrepancies which result in a scattering of the flux reversals on the tape when the current in inverted in perfect synchronism at all heads. Those deviations of the boundaries between magnetic domains from their theoretical positions are the tape skew. The skew is characterized by the maximum distance, measured along the axis, between the first and the last boundaries when a character made of ones only is recorded.

Telemetry System—In telemetry systems, control signals and seismic data are transmitted back and forth via a telemetry line, either hard wire, optical fiber or radio. The data acquisition units (DAVs) spaced out along the geophone lines perform analog amplification, filtering, A/D conversion, control and telemetry functions for one and more channels. The DAVs transmit digital data down the transmission line to a central control unit for recording.

Vertical Seismic Profiling (VSP)—The fundamental goal of vertical seismic profiling is to record and analyze the total seismic signal as a function of depth in the borehole. By using three orthogonal geophones in VSP survey, both transmitted and reflected P and S waves are recorded. Borehole seismic measurement of the down going wave field can be integrated with the upgoing wave field recorded at the surface to provide a better interpretation and understanding of the seismic response of the earth.

Vibrator—A vibrator consists of a hydraulic ram exerting a force between a large reaction mass and a base plate. The base plate is held in contact with the ground by lifting the carrier vehicle on the base plate before vibrating. This allows the base plate to stay in contact with the ground during its movement. Each vibrator is provided with its own vibrator electronics. The sweep which is generated digitally, is converted to an analog voltage and applied to the first stage of the value system.

Vibrator Controls—The most used surface source is a vibrator. Several vibrators are used simultaneously and they must vibrate in perfect synchronism. Each one is equipped with a sweep generator. At the end of a specified delay after receipt of the fire order over radio, the sweep generator generates the pilot signal and applies it to the vibrator valve control system. The pilot signal is generated first in digital form, as a series of samples. Values of a sine function for a number of angles (32 values between 0° and 90°) are stored in a ROM. Associating these numerical values with a sign and calling them at an appropriate speed, a sampled representation of the sweep is generated. The speed at which the values are clocked out is varied as required by the sweep parameters which are the start frequency, the frequency range and the sweep length. An accelerometer is attached to the base plate and its output is compared to the sweep signal oscillographic.

References

1. Crawford, J.M., Doty, W.E.N. and Lee, M.R., 1960; Continuous signal seismograph; Geophysics, Vol. 25, p. 95-105.
2. Gunningham, A.B., 1979; Some alternate vibrator signals; Geophysics, Vol. 44, pp. 1901-1921.
3. Dennison, A.T., 1953; The design of electromagnetic geophones; Geophysical Prospecting, Vol. 1, pp. 3-28.
4. Khattri, K.N., Mithal, R. and Gaur, V., 1979; Pattern space of seismic anomalies associated with hydrocarbon deposits; Geophysical prospecting, Vol. 27, pp. 339-359.
5. Lindsey, J.P., 1971; Filtering characteristics of multiple seismometer groups, long cables and the amplifier input impedance; IEEE Transactions of Geosciences Electronics, Vol. GE 9, pp. 34-42.
6. Nickolls, H.R., 1961; Coupling explosive energy to rock; Geophysics, Vol. 27, pp. 305, 316.
7. Nooteboom, J.J., 1978; Signature and amplitude of linea air-gun arrays; Geophysical prospecting; Vol. 26. pp. 194-201.
8. Pieuchot, M., 1984; Seismic Instrument; Geophysical Press Limited, London, p. 353.
9. Richard, H., and Pieuchot, M., 1956; Seismic efficiency of explosives; Geophysical Prospectiving, Vol. 4, pp. 167-186.
10. Ricker, N., 1953; The form and laws of propagation of seismic wavelets; Geophysics, Vol. 18, pp.10-40.
11. Safar, M.H., 1976; Efficient design of air-gun arrays; Geophysical Prospecting, Vol. 24, p. 773-787.
12. Toksoz, M.N., Johnston, D.H. and Timur, A., 1979; Attenuation of seismic waves in dry and saturated rocks, Parts I & II; Geophysics, Vol. 44, pp. 687-711.

Annexure–A

Geophysical Field System

A Geophysical Field System receives analog signals at its inputs and records a digital representation of those signals as an output. The front part processes analog signals and is made of a number of identical channels in parallel. The number of channels is equal to the number of traces in the spread. The responses of these channels must be identical. This is obtained through careful design and manufacturing. It also requires frequent checking and adjustment. It is preferable to apply filtering to analog data for reducing the noise before applying the automatic gain and for reducing the aliasing frequencies before sampling.

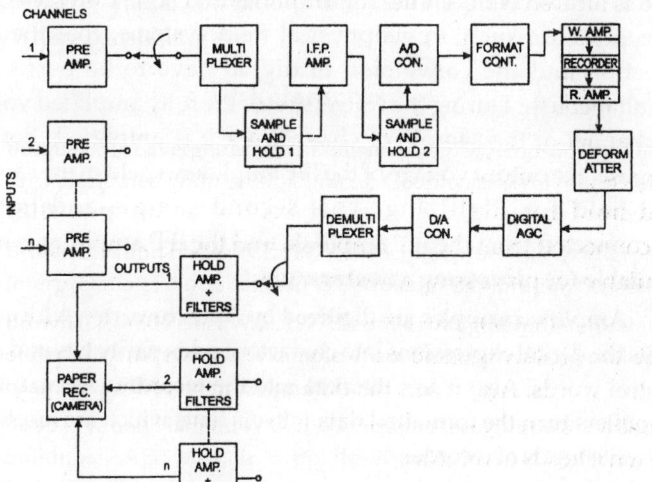

Fig. 1. Block diagram of a simple digital filed recording system.

Consequently, the front end of a seismic data acquisition system consists of a number of preamplifier-filters equal to the number of seismic channels. The preamplifiers are followed by a multiplexer where the input signals are sampled into short time slices which are intermingled and queued through a serial output channel. Samples are held in a first sample-and- hold amplifier. Each sample is amplified to a proper voltages for accurate recording by an IFP amplifier.

The technology available earlier did not allow to sample-and-hold a voltage taken from a wide range of values without introducing unacceptable errors. At that time, the sample-and-hold was located between the IFP amplifier and A/D Converter. Now sample and holds have a higher performance and can accept a wide range of voltages. They can operate on the out-puts of the preamplifiers for instance, before full amplification by the IFP amplifier. Consequently, the first sample-and-hold is located ahead of IFP amplifier, at the output of the multiplexer, and the voltage presented to the IFP amplifier is a fixed voltage, requiring no anticipation of what it will be at the sampling time.

In modern geophysical field systems, a second sample-and-hold is located between the IFP amplifier and A/D Converter. The data rates are such, in geophysical field systems, that the gain selection and the conversion to digital have to be performed simultaneously. During the conversion of the fully amplified voltage of channel n, the sample of channel n + 1 is amplified. For this purpose, the output voltage of the IFP amplifier is held in the sample-and-hold for digitizing. This second sample-and-hold is disconnected from the IFP amplifier, and the IFP amplifier is made available for processing a next sample.

Amplified samples are digitized by A/D converter. A formatter splits the digital expressions into characters, adds parity bits and other control words. Also it sets the data into the recording format. Write amplifiers turn the formatted data into currents which are applied to the write heads of recorder.

All the geophysical field systems include playback facilities. These transform the recorded data back into analog signals which are suitable for a camera display. The read amplifiers restore the bit representations character per character. Data words are reassembled in a deformatter. Since the input voltages extend over a dynamic range much wider than that of a camera display. Therefore, the gain control is performed on digital data by digital AGC. Data words are converted back into voltages by D/A Converter. These voltages are distributed to different channels by a demultiplexer.

The data coming out of the demultiplexer in each channel is made up of discrete voltage samples. It has to be turned into a smooth, continuous, and analog signal. This is obtained by storing the discrete voltages on a capacitor and by applying a ripple filter to the staircase voltage stored in this capacitor. It is usually followed by playback filters applied to enhance the data quality on the monitor record by eliminating unwanted noises.

Note: Different types of geophysical field systems are available in the world market. These different systems have minor changes but the basic principle is the same.

Reference

1. Pieuchot Maurice, *Seismic Instrumentation*, Geophysical Press Limited, London, 1984.

Annexure-B

Chronological Development of Technologies in Geophysical Exploration

In a world where technology is advancing at a rapid rate, it would be highly desirable to have an assessment of future events so that one could plan company business more efficiently. The better this assessment, the better the company's ability to cope, and to capitalize on future events. The Delphi forecasting method is based on expert opinion and is thought to be superior to single expert predictions or forecasts from expert committees.

It is interesting to think about the march of progress in geophysical exploration. What it works best, geophysics grows by accretion of knowledge. Our current state of knowledge is much advanced, but it is often difficult to recall how we got to this point. As with any history, it seems obvious and hardly worth tabulating, if you were someone who lived through it and participated in the development. The value goes to newcomers in the field, those too young to have seem it first hand, and future generations.

But the world changes quickly, and the space of geophysical development is accelerating. In an attempt to keep up, a chronological development of technologies in geophysical exploration is given in below:

Table-1
Chronological Development of Technologies in Geophysical Exploration

Year	*Description*	*Category*
1914	Ministrops Mechanical Seismograph	Acquisition
1917	Fessenden Patent on Seismic method	Acquisition
1921	Seismic reflection work by Geological Engineering Company	Acquisition
1923	D and S Petty measured 2 events, 1 through and 1 around salt dome	Interpretation

Year	*Description*	*Category*
1925	Refraction for salt domes and other high velocity structures (Mexican reefs)	Acquisition
1925	"Interpretation" by spying on where other crews working	Interpretation
1926	Refractions correlated for dip studies	Acquisition
1931	(a) Use of uphole geophone (b) Truck mounted drill	Acquisition
1932	(a) Automatic gain control (b) Interchangeable analog filters	Processing
1933	Use of multiple geophones per group	Acquisition
1936	Rieber Sonograph, first reproducible recording	Acquisition
1939	Started using "Close loops" to check misties	Interpretation
1942	Record sections	Processing
1942	Spliced paper records	Interpretation
1944	Large Scale marine surveying	Acquisition
1950	Harry Maynes experiments with CDP method	Acquisition
1950	Human "Computer" picked records in field and hand plotted cross–sectioning	Inerpretation
1951–56	MIT Geophysical Analysis Group started digital processing and time series analysis	Processing
1952	Analog magnetic recording	Acquisition
1953	(a) First Vibroseis recording (b) Weight–drop recording	Acquisition
1953	Analog Vibroseis correlation	Processing
1954	Gulf, Shell and V. Reynolds removed NMO & static shift and re corded seismic data in variable density.	Interpretation
1956	Central data processing Analog mag tape filtering, mixing (stacking), and record sections	Acquisition
1960	Interpretation Centralized and moved out of field	Interpretation
1961	Analog deconvolution and velocity filtering	Processing
		Contd.....

Year	*Description*	*Category*
1963	Digital data recording	Acquisition
1965	Airgun seismic sources	Acquisition
1967	Depth controllers on marine streamers	Acquisition
1968	Maturing of digital seismic processing packages	Processing
1969	(a) Shear wave crews (b) Introduction of VSP (c) 48-49 channel DFS-III standard recording system	Acquisition
1970	2D migration for spatial reorientation	Processing
1972	Shell, Mobil and Russians used Bright spot or direct hydrocarbon indicators	Interpretation
1973	Anstey, batch display of colour sections based on complex trace analysis	Interpretation
1974	Sam Allan showed sign-bit recording useable	Acquisition
1975	Chevron, Exxon and Conoco seismic stratigraphy, seminar	Interpretation
1976	Gulf and GSI 3D seismic survey using telemetry systems	Acquisition
1976	3D migration for true "focusing" of seismic events	Processing
1977	Gus manufacturing introduced real-time Demux system	Acquisition
1977	(a) Lindseth–seismic log inversion (b) Reconstruction reflectivity equation for acoustic impedance	Interpretation
1979	Interactive processing parameter selection	Processing
1979	Geosource Geostar data base management system.	Interpretation
1980	Digicon, SSC, Litton VAX processing system	Interpretation
1981	Interactive computer graphics for seismic interpretation	Interpretation
1982	(a) Digicon, Secure satellite transmission of marine seismic data during collection (b) Geosource, interactive 3D acquisition design.	Acquisition
		Contd.....

Year	*Description*	*Category*
1982	Cyber-200 and Cray-1 Class victor process introduced to sesmic processing shops	Processing
1982	Colour computer graphics display seismic trace attributes	Interpretation
1983	Interactive 3d Interpretation work stations	Interpretation
1984	2D reflection tomography	Processing
1984	Dip movement	Processing
1984	Optimization theory	Processing
1985	3-D Poststack depth migration	Processing
1985	AVO Analysis	Interpretation
1986	3D reflection tomography	Processing
1986	Bottom cable	Acquisition
1986	Crosswell Seismic	Acquisition
1986	FX deconvolution	Processing
1989	2D attributes	Interpretation
1989	3D prestack depth migration	Processing
1989	Time lapse 3D seismic	Aquisiton
1989	Netural networks	Processing
1990	Subsalt imaging	Interpretation
1991	Wavelet transform processing	Processing
1992	Seismic sequence stratigraphy	Interpretation
1994	Flow parameters—Seismic parameter	Interpretation
1994	Routine anisotropic processing	Processing
1995	3D Attributes (Cherence)	Interpretation
1995	Vertical Cable	Acquisition
1995	Fluid Substitution	Interpretation
1996	Extended Suprious Superphone SM-24	Acquisition
1997	4C multicomponent sea bottom secuiners (node)	Acquisition
1998	GEOX system ARAM24 Seismic	Acquisition
1999	GEOX ARAM24 Vibroseis Signature	Acquisition
2000	Schlumberger Q–land, 30,000 channels and single geophone per trace with 5m group interval	Acquisition

Annexure-C

Technical Terms Used in Geophysical Field System

Authority File— A set of records that establishes authoritative forms for headings or access points for a set bibliographic records.

Asynchronous Transmission—A mode of telecommunication transmission in which each character or small block of data is synchronized. The gap between transmitted characters may be of unequal length.

Alogorithm—A step by step procedure for carrying out a numerical or algebraic operation.

Analog—A continuous physical variable which bears a direct relationship to another variable so that one is proportional to the other.

Band Width—A measure of frequency use or capacity. Voice transmission requires a bandwidth of 3 to 4 KHz.

Batch Processing—A type of computer processing in which similar operations or problems are stored and put in batches to be processed.

Boolean Logic—A form of symbolic logic, devised by George Boole (1815-64) which provides a mathematical procedure for manipulating logical relationships in symbolic form.

Bearing—The horizontal direction of one point with respect to another, usually measured as a clockwise angle. True bearing is with respect to true north.

Binary—Composed of only two elements. A number system in which two digits, 0 and 1 are used.

Binary Gain—A gain control system in which amplification is changed only in discrete steps by factors of 2.

Bit—A binary digit, the smallest unit of information.

Bench-Mark Problem—A problem used to evaluate the performance of computers relative to each other. A flow chart that is coded for execution on various system is often used so that the problem-solving ability of different systems, can be compared under similar circumstances.

Complier—A program for converting a source programme in symbolic language to an object programme in machine language independent of the running of the source programme. An interpreter involves the conversion concurrently with running the programme.

Compatible software—The ability of software to run on different computers without modification. Also, a feature that enables different application programmes, such as Frame work and Word Star, to share common conversions and rules so that they can be run together as one coordinated package.

Computer— A device capable of solving problems or manipulating data by accepting data, performing prescribed operations on the data and supplying the results of these operations. Various type of computers are : analog and digital.

Communication—Transfer of ideas and information from one mind to another mind is called the communication. Man is a communicator of information. It is this capability of communication of very complex ideas from man to man which has led to our present state of civilization.

Cordless Phone—It consists of a base unit and a handset. The handset can be used only within a specified range, upto a maximum distance of 100m from the base unit. A cordless phone unit is connected to land lines and underground cables.

Cable Television—A transmission system using a broadband cable (coaxial cable or optical fibre) to deliver video signals directly to television sets, in contrast to over-the-air transmission.

C-Band—Portion of the electromagnetic spectrum in the 4 to 6 GHz range used for satellite communications.

Data—Information represented in a formal manner so that it can be processed, stored, manipulated, and transmitted by computer or other equipments.

Data Base— An automated body of information that is systematically recorded, retrieved or updated.

DBMS—Database Management System is a special kind of a program that permits users to examine and/or rearrange data-bases in special ways.

Disk Drive —A circular, metal plate with magnetic material recorded on both sides continuously rotating for reading or writing by one or more read/write heads mounted on fixed or movable arms.

Digital— Representation of quantities is discrete units. A digital system is one in which the information is contained and manipulated as a series of discrete numbers.

Downlink—An earth station used to receive signals from a satellite; the signal transmitted from the satellite to earth.

Disk Operating System (DOS)—An operating system that uses disks for its secondary storage medium. Typically, the disk operating system regulates space allocation, keeps track of files, saves and retrieves files and manages other control functions associated with disk storage.

Ethernet—A type of LAN based on CSMA/CD technology developed jointly by DEC, Xerox and Intel.

Ethernet Cables and Bus—Run at a much higher speed and send information in packets. Ethernet bus is a coaxial cable which is routed round the station enabling new modes to be added, without cutting the cable but just screwing on a connection box. The speed of communication through this coaxial cable is 1 MBPS.

Electronic Mail—The electronic transmission of letters, messages and memos from one computer to another.

Electronic Spread Sheet—A software package for variable calculations. The screen display consists of an accountants spreadsheet of upto 60 or more columns and over 250 or more rows. Any element in the display grid can be a numeric value, a formula or an arithmetic calculations; any item can relates to and/or affect any other element.

Ernet Mail—It was established by the Department of Electronics during the Seventh Five Year Plan (1986-91) to provide academic and research institutions with electronic mail facilities. It is currently used by DSIR labs, research centres and academic institutions.

Flow Chart —A diagram showing the operations involved in a process. Often used to show a computer programme or computer processing steps and their sequence.

Fourier Transform—The formulas which converts a time function into the frequency domain representation and vice versa.

Floppy Disks—Storage devices in the form of small flexible disks used for random access requirements in controllers and CPUs and as a compact substitute for punched cards. A typical floppy disk provides capacity for about 3,00,000 data bytes. Floppies were originally developed for low cost, low capacity data storage and relatively low data transfer rates.

Facsimile (FAX)—A system of telecommunication to transmit images for reproduction on hard copy. The original image is scanned and converted to an electrical signal and the electrical signal is subsequently converted to a replice of the original image at the receiving terminal.

Gateway—A protocol translating interface between a given LAN and the external communications environment which uses a different protocol.

Global System for Mobile Communications (GSM)—It is a Pan-European standard and is the latest in cellular technology world wide. The new GSM technology used a digital computer language. The digital transmission eliminates background noise and other interferences. GSM's digital system allows a very large number of calls to occur simultaneously, enabling more people to use mobile phones.

Geostationary Satellite—A satellite with an equatorial orbit 22,300 miles (36,000 km) above the earth. The satellite revolve above the earth in the same direction and with the same period as that of the earth's rotation (i.e., 24 hours), and thus appears stationary when viewed from the earth.

Geosynchronous Orbit—Synchronous with the earth. An orbit 22,300 miles above the earth's equator where satellite circle the earth once in 24 hours, thereby appearing stationary to an observer on earth.

Hard Copy—A printed copy of data in human-readable form.

Hardware—Physical computer equipment, e.g., mechanical, magnetic, electrical or electronic devices.

Host—A system or subsystem in a network that performs actual processing operations against a database and with which other network nodes communicate.

Host Computer—A Primary or controlling computer in a network or system.

Hub—A head end for bi-directional (mid split) networks.

Informational Technology—It is the combination of information science, computer technology, communication technology and management science. Information Technology is the brain of hi-tech industry and telecom is the heart. IT, at a basic level, is the conversion of human intelligence and logic into software that propels the computer to give the desired output. Telecom provides the where withal to send the bytes of information soaring across the country or even to far-off countries via satellite uplinks.

Information Theory—Information theory is the body of statistical mathematics, which has developed out of efforts to understand and improve the communication art. The modern theory of information is based largely on the theories of Shannon. The basic principle adopted is that any message which has a high probability of occurrence coveys little information. Most information is conveyed by the least likely message. Thus information theory is based on probability. Meaningful messages can appear by sheer hence.

Inmarsat—The International Maritime Satellite organization, which provides satellites for maritime and aeronautical communication.

Internet—Internet was born in 1969 as Pentagon research and development network. Now, half of the internet domains are outside the United State. Internet is the ast-growing network of computers around the world connecting million of users through a simple telephone line.

Integrated Service Digital Network (ISDN)—An electronic superhighway for transmitting digital signals using internationally accepted standards.

Indonet—It is an integrated information management and distributed data processing facility. It is a commercial network with nodes at Calcutta, Bombay and Madras on IBM 4361. It is connected to other cities like Ahmedabad, Delhi, Pune, Bangalore through MUX/cluster controllers. This network used IBM's System Network Architecture. It is planned as geographically dispersed computer network linking scattered organizations in the country. It will be connected through international gateways to GTE- Telnet and TYMNET computer networks of USA. Also it will be connected to DATEX-P in Germany and TRANSPAC in France and to the other IR networks. It provides access to remote user terminals.

Infotel—It was established by Informatics India Ltd. in 1989. It is an on-line service center with access to 1500 international databases tapped through DIALOG or EASYNET.

Layer—Used to define a discrete level of function with a defined service interface.

Link—A communication path or circuit between two nodes or points.

Long Haul Network—A wide area communication facility ranging from a few kilometres to around the globe.

Laser Printer—Laser printer are the stars of non-impact technology. Using a laster beam to draw characters directly onto a photo sensitive drum, the full pages image is then transferred electrostatically to toner and paper and then the toner is melted into the paper.

Memory—A part of a computing machine in which data can be stored and from which it can later be retrieved. Types of memory include core, magnetic drum, magnetic tape, registers, etc.

Modem—An A/D and D/A conversion device interfacing the data termination equipment and the analog transmission lines.

Mainframe or CPU—The heart of the general purpose computer that controls the interpretation and execution of instructions. Does not include interface, main memory, or peripherals. It also controls input and output units and auxiliary attachments.

Micro-Computer—A complete tiny computing system consisting of hardware and software that usually sells for less than five thound dollars and whose main processing blocks are made of semi-conducting integrated circuits. In function and structure it is somewhat similar to a mini-computer, with the main difference being price, size, speed of execution and computing power. The hardware of a micro-computer consists of a micro-processing unit which is usually assembled on a PC board with memory and auxiliary circuits. Power supply, control console, and cabinet are separate.

Micro-Processor—The central unit of micro-computer that contains the logical elements of manipulating data and performing arithmetic or logical operations on it. A single chip may contain RAM, ROM and PROM memories, clocks and interfaces for memory and I/O devices.

Mini-Computer—A small programmable general purpose computer typically used for dedicated application. The term often refers only to the mainframe, which typically sells for less than $25,000. Usually it is a parallel binary system with 8, 12, 16, 18, 24 or 36 bit word length incorporating semi-conductors or magnetic core memory offering from 4 to 64 K words of storage and a cycle time of 0.2 to 8 micro-second or less.

Mouse—A hand-held device, separate from a keyboard, used to control cursor position on a display screen. As the mouse rolls along a table top, its relating position approximates the position of the cursor. Introduced by Xerox in the STAR 8010 professional work-station in 1981.

Modulation—It is a process by which some characteristics of the information carrier is modified in accordance with the instantaneous value of the input function. Modulation methods serve to improve the transmission reliability.

Multiplexing—The cost of a communication channel is often very high and therefore it is desired to send number of information on the same channel. The technique of transmitting many messages over the same channel is known as multiplexing. In this case, several sources of information are connected to the same transmitter. At the receiving end, the signals are converted back into respective messages.

Mobile Phone—It is a two-way communication device. Through which both, the caller and the receiver can carry out a complete conversation at anytime and at anywhere. The mobile phone service will also give the facility of "short message service". When one does not want to talk, he can simply receive the message on the display screen of his handset.

Noise—Any unwanted signal.

Nyquist Frequency—A frequency associated with sampling which is equal to half the sampling frequency. Frequencies greater than the Nyquist Frequency will alias as lower frequencies from which they will be indistinguishable.

Node—An identifiable point in a design that must be electrically connected to other nodes. A node may or may not be associated with a specific device or topology.

NICNET—The National Informatics Centre (NIC) was established in 1975 for advanced computer-based methodologies, data management techniques, generating specialized manpower and building intra-city and intercity computer networks. NIC established NICNET in 1977. It links four regional nodes at Delhi, Pune, Bhubaneswar and Hyderabad and has established 32 nodes of State and Union Territory levels and 439 nodes at district headquarters. NICNET is based on spread spectrum multiple access and uses C-200 series micro-earth station.

Network Architecture—A network architecture is the pattern that network components make. It comprises the hardware architecture, distributed applications architecture, applications, integration architecture, communications architecture, etc. Each architecture contributes to the development of an integrated environment. The placement of servers, users, clients, hosts and communication hardware in a suitable pattern gives birth to a functional network architecture.

Offline—A system in which peripheral devices operate independent of the central processor.

On line—A system in which peripheral devices are in direct and continuing communication with the central processing unit.

Operating System—Software that controls the operation of a data processing system and that may provide the following services : determine what jobs are running and what parts of the computer system are working on each job at any given time, impose standards and procedures an machine operation, take care of the numerous little details lumped together as house keeping, and invoking standard troubleshooting actions, in case of malfunction. These operating systems are usually very complex and require large amount of machine and disk storage. Sometimes called Supervisor, Executive, Monitor, Master Control Programme, depending on the computer manufacturer.

Optical Fiber—A thin, flexible glass fiber the size of a human hair which will transmit light waves capable of carrying large number of information.

OSI Standards—Open system interconnect is technically promoted by TC 97, the technical committee of ISO and its other related components. Its work is conducted in CCITT. The important characteristics of OSI are that the systems participating in a network should be open, distributed and heterogeneous. OSI undertakes the job through seven layers. The top three layers cater to users of data and information services, the fourth layer controls the transformation of data between a user and the network. The lower three layers cater to specific functions and are called network layer, data link layer and physical layer. OSI offers a directory of standards that are not only necessary for communication but also important for maintaining bibliographic standards.

Packet Switching—A data network as defined by the CCITT X.25 standard, whereby a virtual call is established but individual packets of data may be separately routed across the network.

Personal Computer—A microprocessor based unit with a central memory typically ranging between 64 KB and 1 MB. PCs having 8 or 16 or 32 BPW are available in the market.

Protocol—A rule of conduct, a procedure for ordering the exchange of formatted information packets between correspondents. Protocols are interpreted through hardware and software.

Pert—Programme evaluation and review technique. A time event network analysis system in which the various events in a programme or project are identified, with the planned time for each, and are placed in network showing the relationships of each event to other events.

Pager—It is a one-way communication device. One can only receive messages through it. These one-way messages are very short, and pass on very limited information. To contact the sender, one will have to use the nearest phone available. Every paging service operates on a specific frequency. One has to first subscribe to a paging service and then buy a compatible pager.

Page—In virtual storage systems, a fixed-length block of instruction, data, or both, that can be transferred between real storage and external page storage, typically about 4 K bytes. A programme will be divided into pages in order to minimize the total amount of main memory storage allocation to the programme at any one time. The pages will normally be stored on a fast direct access store and can be moved into main memory by operating system or hardware whenever the instructions of that subdivision need to be performed.

Paging—In virtual storage systems, the process of transferring pages between real storage and external page storage. If a page is not transferred from auxiliary storage until it is actually needed, then paging is said to be done by demand.

Parallel Processing—Processing more than one programme at a time on a parallel basis, where more than one processor is active at one time, as distinguished from multiprocessing where only one processor is active on one programme at a time.

Powerful—Software is considered powerful if it is efficient and provides a wide range of options. In user software, it refers to the user's ability to put these options into action without too much effort. In system software, it denotes the ability to generate complex commands with a minimum of instructions. Hardware is considered powerful if it is faster, larger, and more versatile than comparable machine.

Quantization—The process of approximation of signal by discrete levels is known as amplitude quantization. The sample is then converted into a code group by means of an encoder.

Relay—A system which performs an intermediate function in the communication between two or more end-systems, e.g., a node in a public switched network.

Redundancy—A repetition of information or the insertion of information which is not new.

RS-232-C—A technical specification published by the Electronic Industries Association establishing the interface requirements between modems and terminals or computer.

Repeater—A device inserted at intervals along a circuit to boast, amplify, and/or regenerate the signal being transmitted. A repeater is needed because the quality and strength of the signal decays over distance.

Remote Area Business Message Network (RABMN)—It was established in 1991 by the Department of Telecommunication. This is a satellite based network with a star configuration and has a master earth station at Sikandrabad near Delhi. This network is used to exchange of messages among closed user groups.

Synergetic—A combination of data elements such that more information is apparent from the combination than from the elements treated independently.

Security—The general subject of making sure the computerized data and programme files of the company can not be accessed, obtained or modified by unauthorized personnel, and can not be fouled up by the computer or its programmes. Security is implemented by special software, special hardware, and the computer's operating procedure.

Software—A term coined to contrast computer programme with the hardware of computer system. Software programmes are stored sets of instructions which govern the operation of a computer system and make the hardware run. Software is a key determining factor in getting more computer power per dollar. Software is used to define the processor programmes, library routines, manuals and other service programme supplied by a computer manufacturer to facilitate the use of a computer. In addition, it may refer to other porgrammes specially developed to fit the users need. All the documents associated with a computer.

State-of-the-art—The most current research and upto data technology in specific field of knowledge.

Sampling—Sampling process converts a continuous signal into a train of pulses. Sampling of data is one of the fundamental operations in digital systems. In a digital system only certain discrete values of amplitudes of the sample are allowed. Any sample which falls between the two allowed levels is converted into a sample of amplitude level nearest to the actual value of the signal.

Sampling Theorem—It states that no loss of information occurs if the signal is sampled at a rate which is at least twice as fast as the signal's highest frequency content. The sampling frequency is determined by the highest frequency content of the signal.

S-Band—Frequencies in the 2.5 GHz range, used for community reception from satellites.

Spread-spectrum Technique—A means of spreading a transmitted signal over extra bandwidth to permit redundancy in transmission and reception without interference from other signals using similar frequencies.

Scientific and Industrial Research Network (SIRNET) —It is a computer communication network initially started for CSIR Laboratories. The SIRNET electronic mail mode is at INSDOC. It is connected to Education and Research Network (ERNET) set up by the Department of Electronic (DOE). Through ERNET,SIRNET has access to Unix User Network, an international network. Also it has access to JANET, CSNET and BINET. The software used for SIRNET is Basic Network Software with standard UNIX package to communicate with other UNIX systems.

Standardization—To promote fast and fault free communication from a source to a user or a user to source, it is important that the conversion of information into signals and signals into information is achieved in efficient way. It is important that international standards must be maintained for switching/packet switching or transmission equipment or protocols.

Telemetering —The transmission of data from a point of observation to a recording point, often by radio.

Transducer—A device which converts one form of energy into another.

Transponder—A device which transmits a signal upon receiving another signals. A microwave receiver, amplifier, and transmitter in a satellite that amplifies and changes the frequency of a signal from an earth station and retransmits it to earth.

Transmitter—The device that sends the telecommunication signal through a wire or cable or through the air.

TAG—A character or digit or a series of characters or digits which are attached to a field or a record as a means to identify or to signal the computer of what is coming next.

Tele Tax—The international service for text interchange which provides a rapid exchange of text via the telephone network and other public data networks.

Terminal—A device for entering data into or receiving data from a computer system or computer network. A point in a communication network at which data can either enter or leaving.

Thesaurus—A terminology list to control access points in a file. Generally used with respect to control of subject terms for use in information retrieval systems.

User Friendly— Refers to software and hardware systems that are supposedly easy for a user to learn without requiring a great deal of specialized knowledge or training.

Uplink— An earth station used to transmit to a satellite; the transmission from the earth to the satellite.

Video-text—The generic name for a system that transmits alpha-numeric and graphic information for display on a video monitor.

Vast—Very small aperture terminal.

Value Added Services—They include facilities such as roaming, teleconferencing, closed user group facility, short message service facility, voice mail, call barring, call forwarding, call hold and call waiting, call number identification presentation or restriction, bill on demand with prior request, advice of charge on handset display screen and many more.

Vikram—It is packed-switched public data network of the DOT. The network management centre is in Delhi, and switching nodes at Bombay, Calcutta, Madras, Bangalore, Hyderabad, Ahmedabad and Pune with 12 remote access point all connected by leased or dial-up lines.

Virtual Terminal—One of the protocols being developed for the OSI application layer for terminal access to computer.

White noise—Random energy containing all frequency components in equal proportions within the bandwidth, phases are random.

Winchester Disk—Rigid, non-removable, magnetic oxide coated, random access disk sealed in a filtered enclosure alongwith the read/write heads and head actuator. Heads fly only about 20 micro-inch from disk surface, allowing very dense data storage. Capacity ranges from 5 Mb (325 to 16,000 pages) for the 5 ¼ inch disk and from 6.5 to 635 Mb (1625 to 15,8750 pages) for 8 inch disk, calculated at 4000 characters per single spaced page.

Windo—An isolated portion of a Cathod ray tube screen that is used to display information independently from the rest of the screen display.

Word Processing—The transformation of ideas and information into a readable form of communication through the management of procedures, equipment and personnel.

Word Star—A popular word processing software package developed by Micro PRO international.

X.25—Interface between the data terminal equipment (DTE) and circuit-terminating equipment for terminals operating in the packet mode and connect to public data network by a dedicated circuit.

X-400—Message handling systems : System model service elements.

Yardstick—Standards of performance for evaluating devices or processes.

Zenith —A point directly overhead.

Three

Seismic Stratigraphy

Introduction

Sedimentary rocks normally exist in more or less parallel layers or strata. Different rock types are distinguished by different physical features. Determination of relative sea level is one of the features which seismic stratigraphy seeks to ascertain from seismic data. The deposition of successive layers of sand and gravel in the bed of a river or in a canal affords a perfect illustration both of the form and origin of stratification. A large portion of the masses constituting the earth's crust are thus stratified. The successive strata of a given rock preserves a general parallelism to each other. But the planes of stratification are not perfectly parallel throughout.

Stratigraphy is the branch of geology that deals with the definition and description of major and minor natural divisions of rocks, and with the interpretation of their significance in geologic history. It deals specifically with the geologic study of the form, arrangement, geographic distribution, chronologic succession, classification, and especially correlation and mutual relationships of rock strata. Also it deals in term of rocks origin, occurrence, environment, thickness, lithology, composition, fossil content, age, history, paleogeographic condition, relation to organic evolution, and relation to other geologic concepts. Stratigraphy thus embraces most of what one might consider as the geology of sedimentary rocks.

Seismic reflectors carry information about the interfaces between rock types in the earth. Lateral changes in rock type are in general rule, although usually rock type changes occur gradually.

For example, sediments deposited near their sources are generally coarser grained than those deposited further away, and we expect a formation that is mainly sandstone to gradually change into a shale as we move away from the source. It is also expected that other types of formations to change their nature along the bedding. Changes in rock type produce changes in the reflectivity, which affect the wave shape seen in seismic data. Inferring stratigraphic changes and where they occur based on characteristics of seismic data is an objective of seismic stratigraphy.

Data recorded at sea are usually more uniform than land data because the recording environment is more homogeneous and produces fewer distortions, so that it is easier to see variations attributable to stratigraphic changes. Bright spots are now regarded as a subset of seismic stratigraphy. The most obvious information obtainable from seismic data is the configuration of interfaces, and the historical use of seismic data has been in the mapping of geologic structure. Most stratigraphic conclusion from seismic data have been inferences from the structure that the seismic data determine. The distinction of seismic stratigraphy is that it goes beyond merely mapping structure and searches for more direct seismic evidence as to the nature of the rocks and the fluid contained within their pore spaces.

Detailed Encyclopaedia

This encyclopedia is arranged in alphabetical order. The detailed encyclopaedia is given below : –

Constant Time Indicators (Seismic Reflections) – Seismic reflections generally indicate isochronous surfaces or unconformities. A constant time surface at some-time was the surface of the solid earth. A time surface does not necessarily indicate the same lithology contrast everywhere. Different facies may have been deposited above or below different portions of the surface. Seismic reflections generally indicate time surfaces rather than the attitude of seismic facies lines, where they differ in attitude. The explanation is partly geological, partly geophysical, and partly a matter of sampling density. Geologic events are often episodic, i.e., some event such as a storm or flood rearranges the sediments over a large area in a short period of time. The long period of steady deposition between these relatively sudden events may provide the sediment materials, but the sudden events rearrange them and determine the final pattern. These episodic events are often widespread, and the surface that they mark indicates constant time. A succession of such events produces a number of nearly parallel minor but widespread markers.

Seismic reflections usually are the interference composites of many sub-reflections. Sub-reflections result from the often-minor contacts along the time lines. One does not see the individual sub-reflections because of resolution limitations, but they tend to add in the same way at geophone stations that are not very far apart because the pattern change is so small between adjacent reflecting points. We pick a seismic reflection by deciding what is the coherent from trace-to-trace, and thus we tend to pick as reflections the attitude of the time lines. Facies boundaries are often very irregular or gradational and hence do not produce coherent reflections over appreciable distance. However, when a facies boundary is fairly sharp and widespread, it does, of course, produce a reflection.

Convolution Model—In stratigraphic modeling we are usually concerned with a distribution of rocks with different physical properties, and our objective is to see how this distribution affects the passage of seismic waves. The convolution model states that a seismic trace is the result of convolving the reflectivity function of the earth with a wavelet, with noise added. The reflectivity function is the waveshape we would record from the actual earth if the wave let were impulsive, and the wavelet, with noise added. The reflectivity function is the waveshape we would record from the actual earth if the wavelet were impulsive, and the wavelet is the waveshape from a single reflector using the actual wavelet.

The physical picture that the convolutional model implies is of a downgoing wavetrain, the wavelet, which is successively reflected from places at which the physical properties change. The magnitude of a reflected wavelet is proportional to the magnitude of the change in acoustic impedance (the product of density and velocity). The polarity of the reflected wavelet depends on the sign of the change.

The superposition of all the individual reflections plus noise gives the seismic trace. When the angle of incidence at an interface is perpendicular (normal incidence), the ratio of the amplitude of the reflected wave compared to the amplitude of the incident wave is called the reflection coefficient or reflectivity. Most seismic reflection work involves angles of incidence that are sufficiently close to zero within about ± 20 degrees. This relationship gives a close approximation. The earth's reflectivity is simply the sum of the reflectivities of all the interfaces in the earth, separated by the two-way travel time between them.

Defining Reservoir Limits– Seismic data can be especially valuable in extending well information to the surrounding region. Many uncertainties are reduced significantly by well control. One application of hydrocarbon indicator analysis is in delineating the limits of reservoirs for use in field development. Attributes can show faults and other features that separate different pools and aid in mapping the area extent of various pools. The gas reservoirs are associated amplitude and low frequency anomalies when the reservoirs exceed about 10 feet in thickness. Faults that are seen clearly in the seismic data separate the field into a number separate pools. The limits of the pools as determined by the seismic data and by subsequent drilling are normally in good agreement. Oil producibility limits are apparently controlled in different places by porosity development, gradual water saturation variations, or an oil/ water contact. While the seismic data do not every where predict the oil limits correctly, but seismic data assist in the field development.

Fresnel-Zone Effects—A factor in horizontal resolving power is the Fresnel-zone. It is the portion of the reflecting surface from which energy returns to a geophone within a half-cycle after the reflection onset. Energy reflected from this zone interferes constructively and builds up the reflection. The reflections from reflectors of different lengths are measured in terms of the Fresnel-zone size. Where the reflector is larger than about one Fresnel zone, the reflection shows the shape of the reflector where as for small reflectors, the arrival time patterns are almost identical. But there are

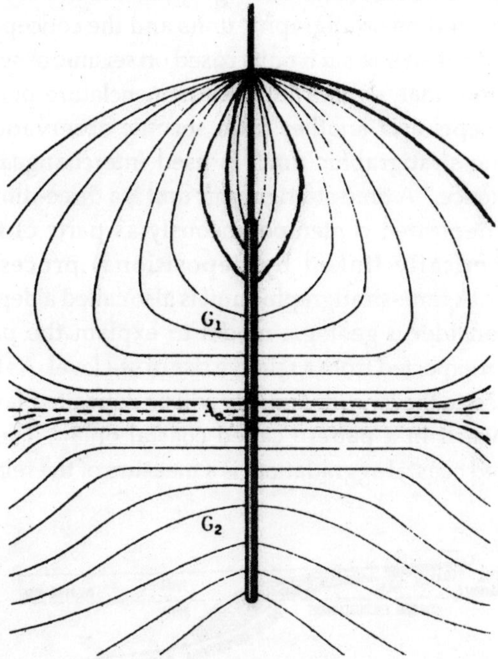

Fig.1. Fresnel zone for spherical waves (a) First Fresnel zone (b) Low frequency Fresnel zone is larger than high frequency ones (After Sheriff, 1977a).

amplitude and other differences. Fresnel-zone effects result from variations along the line of the seismic section without occurring off to the side, i.e., horizontal resolution can be improved by migration. However, while migration removes Fresenel zone smearing, it substitutes migration noise and it does not correct for-out-of-the-plane effects. Spatial sampling, i.e., the distance between geophone groups in seismic data acquisition and the areal extent of geophone groups, also limits horizontal resolution. The cental points of refletor sampling are spaced at half the geophone group spacing and differences between these successive samplings have to indicate the reflector attitude. Where spatial sampling is too coarse, ambiguities in dip result. This effect is called spatial aliasing.

Geologic Models—The stratigraphic model for seismic analysis is one of time-stratigraphic units and the concept that the section can be broken into such units based on seismic observations. Seismic sequence analysis uses duplicate nomenclature, one set form geologic concepts and another form seismic observations, e.g., geologic "time stratigraphic unit" is used interchangeably with "seismic sequence." A time-stratigraphic unit is a three-dimensional set of facies deposited contemporaneously as parts of the same system, genetically linked by depositional processes and environments. A time-stratigraphic unit is also called a depositional sequence. Consider a geologic model to explain the pattern of sedimentation expected from a relative rise of sea level. As the water rises compared to the land, sediments will be deposited farther and farther landward in a pattern called coastal onlap. The vertical distance, called coastal aggradation, is a measure of the relative rise,

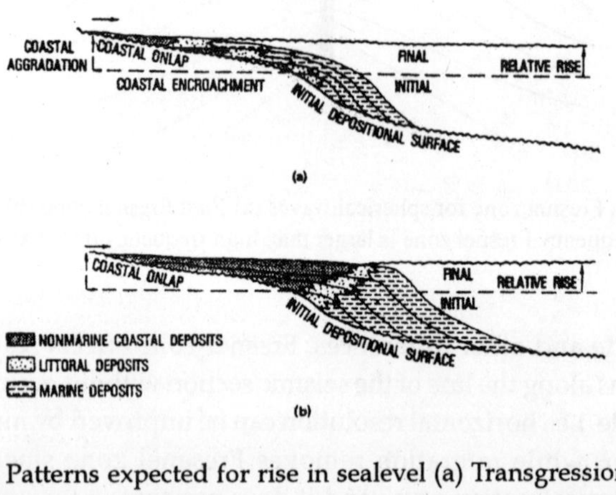

Fig.2. Patterns expected for rise in sealevel (a) Transgressionn (b) Regression (After Vail, Mitchum and Thompson, 1977a).

allowing for error because of rotation of the overall section and other factors. The associated horizontal distance is called coastal encroachment. The nature of the facies that will be deposited at any location depends on both the amount of sea level rise and the

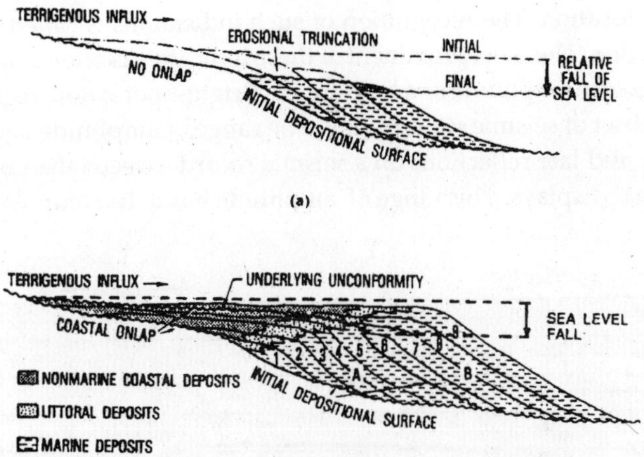

Fig. 3. Patterns expected for fall in sealevel (a) A gradual sealevel fall (b) A rapid fall in sealevel (After Vail, Mitchum, and Thompson, 1977a).

availability of material for deposition. If the terrigenous influx is low, the coast line will move landward with time, producing a transgression. However, if the terrigenous influx is especially great, the coast line may move seaward with time despite the sea level rise.

In a gradual fall of sea level, successive deposition would be displaced seaward. The tops of the patterns would subsequently be eroded off as they were exposed. If a period of rising sea level should be followed by a sudden fall, we expect the entire depositional pattern to shift seaward by a considerable amount, giving the pattern indicated in this figure. This picture is one of asymmetric depositional cycles. The pattern of gradual rise followed by sudden sea level fall has presumably been repeated a number of times. These patterns in different parts of the world are contemporaneous. This has led to the eustatic level concept that periods of rising and falling sea level represent the same geologic time world-wide. Eustatic chart reveal the same type of pattern repeated on different scales, which has led to the concepts of cycles within cycles.

Hydrocarbon Indicators—Hydrocarbon accumulations sometimes have effects on seismic data that can be used to indicate their location. The recognition of such indications is called direct detection. The most prominent of these indicators is often a marked increase in amplitude or a bright spot. Bright-spot is now regarded as subset of seismic stratigraphy. The range of amplitude between early and late reflections on a seismic record exceeds the range of normal displays. This range of amplitude has to be compressed to

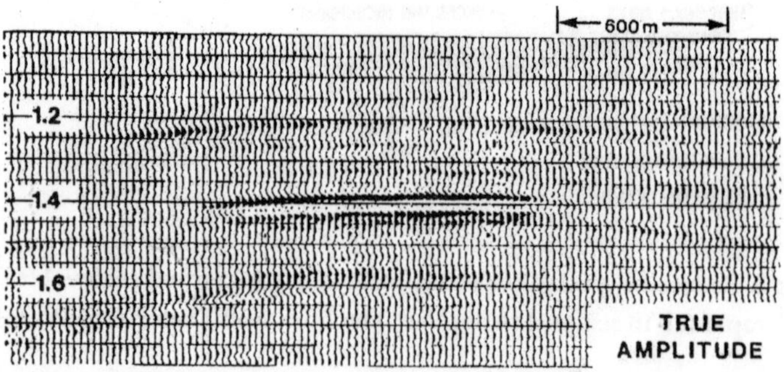

Fig. 4. A hydrocarbon accumulation indicated by high amplitude event (After Schramm, Dedman, and Lindsey 1977).

permit both shallow and deep reflections to be seen. Automatic gain control was used earlier for amplitude compression. Bright-spot technology has shown that hydrocarbon accumulations, especially in Tertiary clastic sediments, are often associated with amplitude anomalies. The presence of hydrocarbons in the pore space of a rock lowers the velocity and also the density. In Tertiary clastic sediments the lowering of the velocity and density is often so large as the produce exceptionally large acoustic-impedance contrasts, resulting in high-amplitude reflections. In other types of sediments, hydrocarbon accumulations may produce sufficient changes that variations in reflectivity along the bedding evidence the accumulations. However, acoustic impedance can change for a number of reasons so that no unique relation exists between amplitude and accumulations.

The low velocity of a hydrocarbon accumulation produces a slight delay in energy that passes through the reservoir and sometimes a velocity sag can be seen in deeper events. Although the delay has to have been produced by a thick gas accumulation. Other types of velocity effects are occasionally seen also, such as distortion, because of the lensing effect on ray paths that pass through a gas reservoir. The reflectivity of a bright spot may be sufficiently great to make it an important generator of multiples. A section processed with constant stacking velocity, the stacking velocity for the bright spot, emphasizes multiples and provides a useful tool for locating the accumulation that generates the family of multiples. A lowering of amplitudes is often seen underneath accumulations. This is sometimes attributed to loss of energy in transmission through the accumulation, but the effect more often results from amplitude normalization. The amplitude of events above the accumulation thus may show weakening over an accumulation as well as those below the accumulation.

Probably the most reliable indicator is the flat spot. A flat reflection presumably resulting from a gas/oil, gas/water or oil/water interface within a reservoir. However, flat spots are usually not seen distinctly unless the reservoir is more than a quarter wavelength thick, so that different reflections associated with the reservoir are at least partially resolved. The lowering of velocity in a hydrocarbon accumulation sometimes changes the reflection coefficient at the top of a reservoir from positive to negative, resulting in polarity reversal. More commonly, the reflection from the accumulation is only one component of a composite reflection so that a change in this component changes the phase of the composite. Sometimes the lowering of velocity result in the contrast against capping sediments becoming smaller, so that accumulation is evidenced by a weakening of the amplitude, producing a dim spot.

No hydrocarbon indicator by itself is foolproof. A conjunction of indicators provides a much reliable indication of an accumulation than any individual indicator. Thus, strong amplitude with a low-frequency shadow at a location where trapping conditions might exist, e.g., in the crest of an anticline or with a fault at the updip edge, can together provide a strong case. If there should be flat spot in addition, the case is very strong.

While no single attribute provides a reliable indicator of hydrocarbons by itself, a conjunction of some of the following indicators provides strong evidence—

1. Amplitude change gives a brightening or dimming of the reflection amplitude. Amplitude also affects the generation of multiples because of the change in the reflection coefficient.
2. Frequency change, generally involves a lowering of frequency immediately below the reservoir.
3. Velocity change, almost always involves a lowering of velocity. Where the reservoir is sufficiently thick, it may produce a velocity anomaly in reflection beneath it.
4. A change in wave shape sometimes shows cup as a reversal of polarity and sometimes by other phasing changes.
5. A flat spot can result when a reservoir is sufficiently thick.
6. Various other things are :- (1) association with a trapping mechanism, e.g., reservoir is being located on an anticline crest or at a fault, (2) a presence of associated accumulation, (3) indications of gas leaking out of a reservoir, etc.

Mapping Seismic Sequences (Three Dimensions)—The sequency boundaries have to be followed over a grid of intersecting lines to produce a three-dimensional map of the sequence unit. The seismic evidence may be somewhat different on lines in different directions, e.g., a dip line across a fan may see progradational patterns whereas a strike line may show overlapping cusps.

Fig. 5. Seismic facies patterns (After Sheriff, 1980).

A map can be made for each sequence unit showing the geographical distribution, thickness and shape of the unit. The nature of the evidences that mark the sequence boundaries is also transcribed to a map. The Exxon convention express reflection character features as a letter fraction, indicating the evidences at the upper and lower boundaries of the sequence above the line and the configuration of the reflections within the sequence below the line. This subdivides the unit into different portions on an objective basis. The different portions of the unit may be subsequently interpreted in terms of facies, e.g., concordant cycles may be interpreted as a delta, the shape of the unit and its relation to other nearly units providing assistance in this identification. Other available information, such as well control, seismic velocity measurements, is incorporated and thus a three-dimensional picture of each time stratigraphic unit is built. The interpretation as to depositional setting, facies, as rock type is based upon ordinary geologic reasoning once the picture of the unit is developed, taking into account the relationship of the unit to neighbouring units.

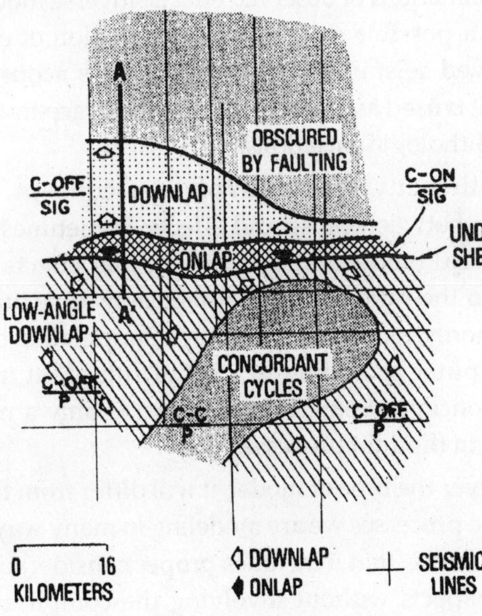

Fig. 6. Map of a seismic sequence showing the evidences for the unit (After Mitchum and Vail, 1977).

Migration—Migration involves repositioning data from a system referenced to the points of observation to a system referenced to the reflecting points. It attempts to position data so as to give an image of reflector. Most stratigraphic evidence, e.g., evidence of uncomformities, onlap, reeafing, etc., involves angularity between two or more reflections, and hence the reflection elements will not be correctly located with rest to each other unless the data are migrated. Isopach maps showing the time intervals between reflection events provide indications for reversal types of stratigraphic features. Significant isopatch variations sometimes involve only a few milliseconds.

Modeling Concept— Modeling is the principal tool used in reflection character analysis. Both forward and inverse modeling are used. Forward or direct modeling involves computing the effects of a model. It starts with a model of the lithology and calculates the seismic trace. It is used to see how stratigraphic changes affect seismic data and to test whether postulated stratigraphic chants provide feasible explanations of observed effects. Inverse modeling involves calculating a possible model from observation of effects. It starts with observed seismic traces and calculates acoustic impedance variations. It is used as an aid in interpreting trace-to-trace variations in terms of lithologic variations.

Modeling invariably involves a concept, an idea of a relationship between cause and effects. Sometimes it involves an actual physical model and the concept that effects will be scaled according to the ratio of model dimensions to actual dimensions. More commonly, modeling is done by manipulation of numbers in a digital computer according to a set of rules that incorporate the modeling concepts. Sometimes a model is only a mental concept that we use in thought processes.

Whatever the type of model, it will differ from the actual earth and from the processes we are modeling in many ways. The essence of a good model is that it includes proper consideration of the most important aspects without involving the complications of lesser important aspects. However, judgment is required to separate important from unimportant considerations. What is important may depend on specific objectives, so that modeling may change as objectives change. Furthermore, there is no assurance of uniqueness in modeling since it is always possible that a different model or different modeling rules may give an equivalent result. Thus, achieving a desired result is not an adequate test of the validity of the model nor of the modeling program. On the other hand, while modeling cannot prove a point because of non-uniqueness, it can disprove, i.e., it can show that a certain model can not represent the situation that gives observed results.

Modeling involves incisive analysis as to what is or is not important, and thus provides one of the best ways to develop and understanding of which conclusions are reasonable and which are beyond our present abilities to distinguish. Its educational value for stratigraphic interpretation is tremendous.

Picking of Unconformities—Reflections generally indicate either (1) time (isochronous or stratal) surface or (2) unconformities. The fact that unconformities are usually good reflectors. This is important because the mapping of unconformities is the key to seismic sequence analysis. Unconformities represent a hiatus in deposition during which conditions are apt to have changed, so that there are different lithologies above and below, thus producing a contrast and a reflection. Furthermore, unconformities are often associated with some angularity, and the angularity of reflections provides the key for picking and identification. Because unconformities are among the best reflectors, we can often follow an unconformity on the basis of its reflection strength even where it is not marked by angularity. The technique for mapping seismic sequences is to first recognize unconformities where angularity (on-lap, erosion and so on) is evident and then follow the unconformities through regions where there is no angularity by virtue of their amplitude standout. The procedure is to mark angularity evidences on the seismic section and then draw in the unconformity considering the location of these evidences and the amplitudes of events. In this way different seismic sequences can be separated.

Picking Seismic Sequences— Presumably, conditions were relatively constant during the alying down of a depositional unit but then changed before the next depositional units was deposited. Since we identify depositional units with seismic sequences, we expect a seismic sequence to gradually onlap previous sediments in the landward direction, have a thickest portion, and thin again in the seaward direction as the amount of sediment decreases because of remoteness from the sediment source. We expect onlap, erosional effects, and so on, at the boundaries between seismic sequences. The angular pattern of reflections at sequence boundaries is the key to isolating sequence units. Of course, the sediments within a unit may be concordant with the underlying and/or overlying boundaries in places and we may have to push or phantom through such regions to complete the bounding surfaces.

The seismic sequence analysis procedure is to pick the unconformities that bound units and thus isolate them. The seismic sequence units are often replotted as a chronographic or chronostratigraphic section, i.e., as a diagram wherein age decreases upward and the lateral extent of units is displayed on the horizontal scale. Such a diagram emphasizes missing portions that have to be explained as missing because of either erosion or non-deposition.

A relative sea level chart is also made, the relative magnitude of a rise being obtained from measurements of the coastal aggradation that can be seen on the seismic section, and the relative amount of a fall being obtained from the difference in level between a unit that is displaced a significant distance seaward compared to the top of the next older unit. The values so obtained may not indicate correctly the amount of sea level rise or fall, but they will at least give general picture. The relative sea level chart should be made as objectively as possible. Adjustments for the effects of compaction, local structure, regional tilt, and so on, should be left until a more comprehensive interpretation stage is reached. Adjustments made piecemeal on individual lines are apt to be made inconsistently. The next step is to correlate the relative sea level chart with a eustatic level chart to age-date the sediments. The key is to correlate first the periods of major sea level fall. These are of limited number, and once they are correlated, the intervening pattern can be worked out.

Reef Patterns—Reefs can be of a number of different types, e.g., patch reefs, shelf-margin reefs, pinnacle reefs and barrier reefs.

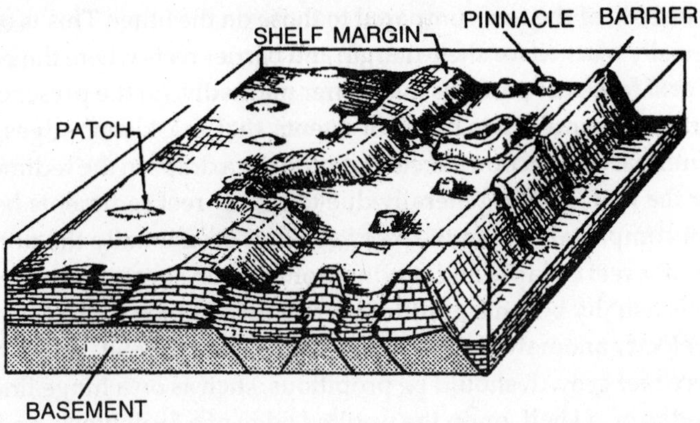

Fig. 7. Types of carbonate build-ups (After Bubb and Hatledid, 1977).

Carbonate platforms and other carbonate sediments are also sometimes considered part of reefs. Reef does not necessary imply a bioherm built by coral or other organisms. Shelf-margin and barrier-type reefs are distinguished by their elongate shape. They usually separate different depositional environments, often a quiet, shallow lagoon from deeper water deposits. They can often be distinguished by different reflection patterns, so that the lateral change in reflection pattern in locating them. The reefs themselves are often so irregular they do not produce coherent reflections, although frequently the top of the reef or sediments draped across the top produce a reflection that outlines the reef. Sometimes reef edges or the termination of reflectors against reefs produce distinguishable diffractions. The shelf edge or fault corner that provides the locale, on which the reef grows can sometimes be seen. Patch and pinnacle types reefs are generally smaller and more difficult to detect on seismic data. The primary key to finding them is knowing where to look, i.e., knowing the portion of the section where conditions of their growth were apt to be propitious.

Criteria for Distinguishing Reefs on Seismic Sections are:—(1) reflections that partially outline reefs, (2) reflection voids, distinguishable by the sharp termination of reflections that onlap the reef, (3) changes in amplitude, frequency, or continuity of reflections at reefs edges, (4) differences between reflection patterns on one side of the reef compared to those on the other. This is often especially marked for shelf-margin and barrier reefs where the back-reef and fore-reef patterns may differ markedly, (5) the presence of diffractions and other types of events that mark reef edges, (6) differential compaction effects that produce a drape in the sediments over the reef. This is generally due to the off-reef sediments being more compactable than the reef itself, but occasionally the porous parts of a reef collapse and produce compaction effects. Compaction effects usually become gradually less with distance above the reef, (7) velocity anomaly for reflections underneath the reef, (8) location where reef growth should be propitious, such as on a hinge line, at the edge of a shelf, or on the uptilted edge of a fault block, and (9) regional factors, such as the knowledge that the climate or environment associated with a particular reflections was propitious for reef growth.

Reefs are usually evidenced by a combination of the foregoing, it being recognized that none are exclusively indicators of reefs. Commonly, a reef has higher velocity than the surrounding sediments, especially where the surrounding sediments are shale's, and hence reflections under a reef usually show an apparent but unreal uplift. Sometimes a reef is surrounded by evaporate deposits such as anhydrite, which have higher velocity than the reef, and then the velocity anomaly under the reef is an apparent sag. Occasionally the back-reef and force-reef sediments will be of sufficiently different kinds that velocity anomalies will differ for opposite sides of a reef.

Reflection Character Analysis—Reflection character analysis involves study of the changes in the waveshape of one or more reflections associated with a reflectivity change. The objective of reflection character analysis is to interpret trace-to-trace waveshape changes in therms of changes in the stratigraphy or fluid in the pore spaces. Reflection character analysis usually involves both forward and inverse modeling. The initial focus of reflection character analysis was on hydrocarbon detection. The properties of a reservoir generally change with the nature of the

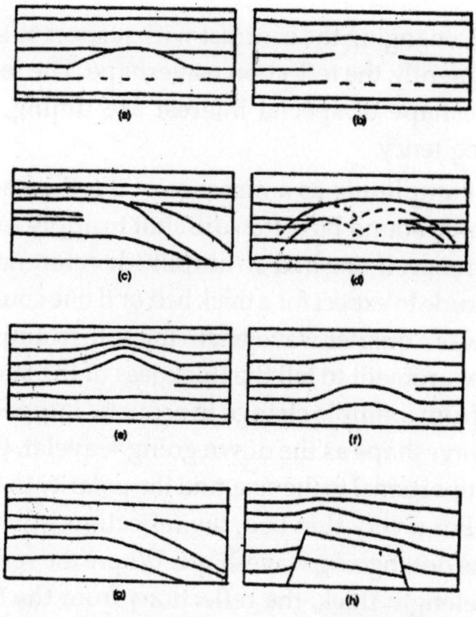

Fig. 8. Criteria for reef identification (a) Reef outlined by reflections (b) Indicated by reflection void (c) Change in reflection pattern on opposite side of reef (d) Diffractions from reef edges (d) Differential compaction over reef (f) Velocity anomaly underneath reef (g) Ree located on hingelina and (h) Reef located on structural uplift (After Bubb and Hatledid, 1977)

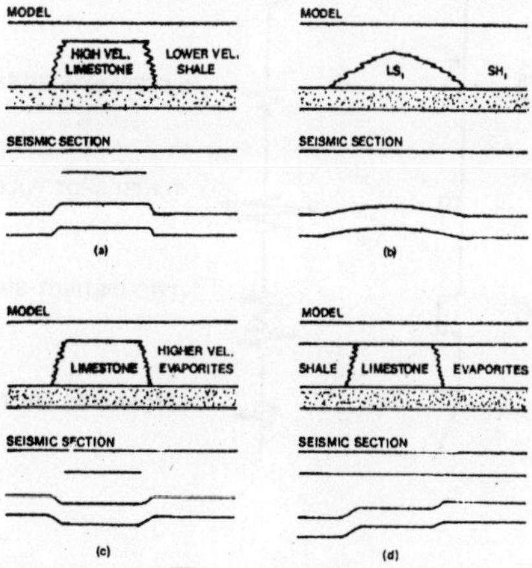

Fig. 9. Velocity anomalies resulting from reefs (After Sheriff, 1980).

interstitial fluid, changing the contrast with respect to the reservoir seal and, consequently, the reflection waveshape. The features of the reflection waveshape of special interest are timing, amplitude, polarity, and frequency.

The use of amplitude as a measure of the thickness of a thin bed is theoretically sound but often difficult to apply for want of an amplitude reference. If one had an amplitude reference, i.e., if one knew the amplitude to expect for a thick bed or if one could determine where the bed was a quarter wavelength thick, then amplitude could be a useful measurement to tell the thickness of the bed when thin. The reflection from a simple change in acoustic impedance (a step) has the same waveshape as the down going wavelet, the reflection amplitude is proportional to the size, and the polarity to the direction, of the change. For a very thin bed, the reflection waveshape is the derivative of the downgoing waveshape. Where the reflecting bed is a quarter wavelength thick, the reflections from the top and base

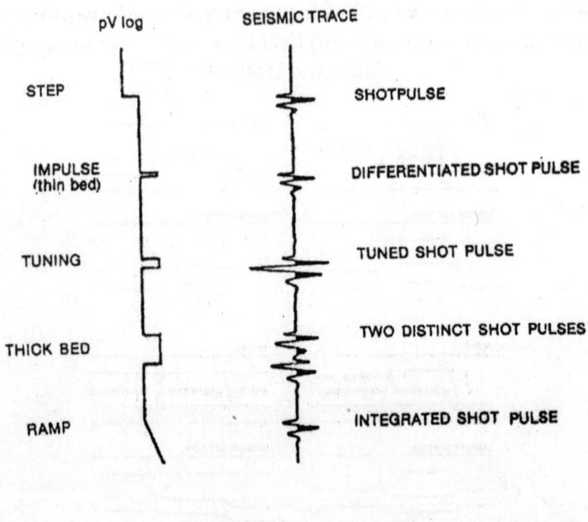

Fig. 10. Reflections from simple interfaces (After Sengbush, Lawrence and McDonal, 1961).

interfere constructively and give exceptionally large amplitude, an effect called tuning. Where the bed is very thick, the reflections from the top and base are completely separated. Where the acoustic impedance change is gradational, the reflection waveform is the integral of the down going waveform. Much modeling is concerned with structural features or with combinations of structural and stratigraphic features. Models are often much more complicated and can involve lithology changes and hydrocarbon traps as well as structural features. A use of modeling is learning how effects appear and whether they are large enough to be seen in actual seismic data.

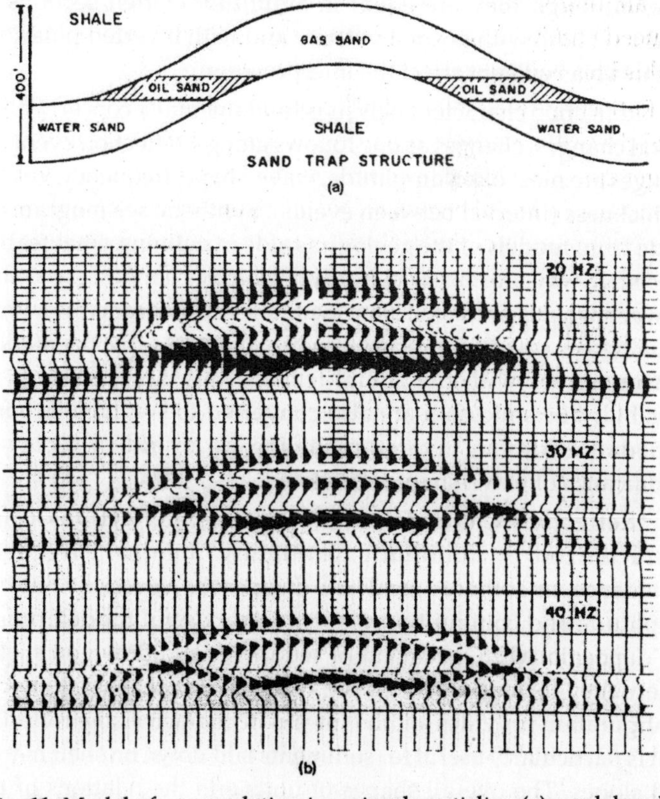

Fig. 11. Model of accumulation in a simple anticline (a) model and (b) Synthetic seismograms for 20, 30 and 40 Hz Ricker wavelets (After Sheriff, 1980).

While models may be quite complicated, they are still very simple compared with the real earth is a model of fairly simple anticline containing gas/oil accumulations. Wavelets with dominant frequencies of 20, 30, and 40 Hz have been used to make synthetic seismograms. The sand thickness is just about a quarter wavelength in thickness in the watersand case for the 20 Hz wavelet. The reflection from the base of the sand appears at first glance to be much stronger than that from the top of the sand, but this is an illusion produced by the variable-area display, which emphasizes peaks by shading them in, so that the trough associated with the sand top does not make as much impact as the peak from the sand base, although they are equal in amplitude. Often sections are produced both with normal polarity and with inverted polarity so that this bias will not affect the interpretation.

Reflection character analysis is fundamentally concerned with lateral changes, changes as one follows along a reflection event. The changes are most often amplitude, wave-shape, frequency, velocity, or thickness (interval between events). Synthetic seismograms are made from models of the geology as aids in defining what features of seismic wave form associate with geology changes. Agreement between what is predicted from synthetic seismograms and what is actually seen on seismic records is not a guarantee that the modeled feature agrees with the actual feature, but merely that is one possibility. Seismic data are also converted to synthetic acoustic-impedance logs in inverse modeling as an aid in seeing the significance of wave-shape variations.

Reflection Characteristics (Types)— Reflections can be characterized by amplitude, continuity, dominant frequency, abundance, smoothness, etc. These distinctions, especially reflection amplitude and continuity, are particularly useful for determining the depositional environment of sediments laid down on shelves, where they form the basis of the classification scheme. Reflection configuration provides another means of characterizing reflections that is particularly useful for sediments laid down on shelf margins and slopes. The overall shapes of units and the relations of their shape to internal reflection configuration is particularly useful for classifying sediments laid down on basin slopes and floors.

The most common types of reflection configurations consist of parallel, sub-parallel, or divergent reflections. Divergent reflections gradually spread out, almost always in the down-dip direction, indicating gradual basin subsidence during deposition, with gradual basinward tilting. New reflections appear between other reflection as units thicken and become resolvable. The bounding configurations of reflections at the base of seismic sequence units are base onlap, downlap, or parallelism to the boundary. The distinction between onlap and down lap is based on the geometry, i.e., whether the on lapping bed at its termination is flat to dipping upward or is dipping downward. Sometimes onlap relates to those portions of the sequence nearest the coast and downlap to those portions at the seaward end of the unit. In the case of onlap, the distinction implies thinning because there was not much room for deposition. In the case of downlap it implies thinning because there was not much sediment available. In the early phases of an interpretation, distinction should be made on the basis of geometry only, without regard for the implied reason.

The bounding configuration of reflections at the top of seismic sequence units are toplap, erosional truncation, or parallelism to the boundary. The distinction between erosional truncation and toplap is whether the sediments did or did not extent significantly higher than their present situation indicates. Toplap implies deposition near the wave base where there was appreciable energy during or shortly after the deposition of the sediments, so that there is reasonable probability that sorting of grain sizes by the wave energy will have occurred. Erosional truncation, on the other hand, implies that the sedimentary unit formally extended significantly beyond its present limits, i.e., possibly being exposed above the water, and that the top of the unit was removed. The distinction is not always clear. Resolution usually limits the detail that can be seen at the boundary. Ringing of an unconformity reflection because of inadequate processing also sometimes obscures what happens. Because ringing gives a wavelet a duration of several cycles.

Patterns within a sequence unit are occasionally very complex and are given a variety of descriptive terms, e.g., hummocky, shingled, contorted, wary, lenticular, disrupted, chaotic, irregular, variable, reflection-free, etc. The most common complex internal patterns indicate progradation. These are divided into two classes, oblique and sigmoid. The sigmoid pattern is distinguished by a very gently "S" shape, with the top to the pattern tending to parallel the sequence boundary, whereas the oblique pattern shows toplap or angularity with the upper boundary of the sequence. The sigmoid pattern represents deposition in quiet (usually deep) water with low depositional energy so that the top of the unit was not disturbed during deposition. The sigmoid pattern therefore is indicative of poor sorting and fine grain sediments. The oblique pattern represents deposition near the wave base in a high-energy environment, and thus sorting of grain sizes may have occurred at the tops of oblique patterns. What happens at the top of progradational patterns is not always clear because of resolution limitations. What appears to be a toplap pattern may indicate merely that the intervals are too thin for the reflection to be resolved. The distinction between oblique toplap and erosional truncation may also be difficult to make. Coastal onlap

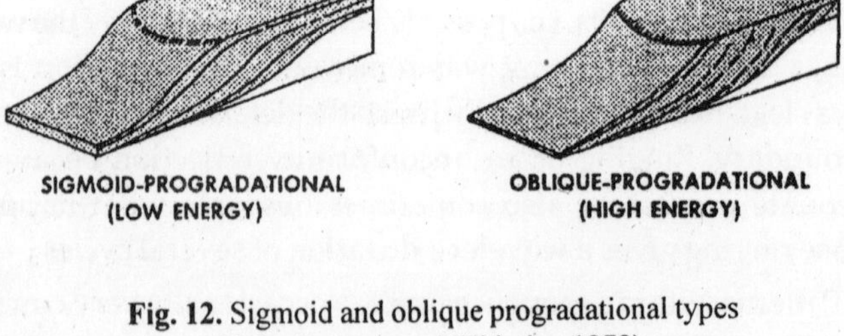

Fig. 12. Sigmoid and oblique progradational types (After Sangree and Widmier, 1979).

implies that sea level was rising with respect to land level. Sediments in a coastal onlap situation were deposited near sea level, probably slightly above sea level, e.g., fluvial deposits. Marine onlap implies deposition well below sea level. The three-dimensional shape of basin-slope and basin fill units often provides the fundamental key to their identification.

Offlapping prograding patterns on the slope, generally indicate a sustained supply of sediment and imply a river source, but they are usually fine-grained because they are so far from the sediment source. Carbonate-shelf systems generally prograde in segmoidal patterns. Onlapping fans on slopes result from turbidity flows derived from the adjacent shelf sediments. They may be of reservoir quality if the shelf sediments contained coarse-grained material. However, widespread onlap probably indicates slow, fine-grained deposition.

Resolution—Resolution is defined as the minimum distance between two features so that one can see the two features clearly. Consider the reflections from a wedge that has a velocity intermediate between the velocities above and below the wedge. When the wedge is thin, the reflection pattern is nearly the same as for a single interface. When the wedge thickness is greater than approximately ¼ wavelength, the reflection pattern shows that at least two interfaces are involved. The minimum thickness of a bed in order to see the effects of the top and the base of the bed as distinctly separate is the resolvable limit and it is about ¼ wavelength. This is applicable to a minimum-phase wavelet. A minimum-phase wavelet has most of its energy concentrated towards the front of the wavelet. Most of the wavelets in actual exploration are nearly minimum phase. The wavelet shape most desirable for interpretation is zero phase. Zero phase wavelet is symmetrical about some central point.

If the velocity above and below a wedge is the same, the reflection pattern is almost indistinguishable from that of a single interface unless the wavelet exceeds ¼ wavelength in thickness. In this case the resolvable limit between a minimum-phase wavelet and zero phase wavelet is the same. The changes in amplitude and their use in determining the thickness of thin beds is important. The

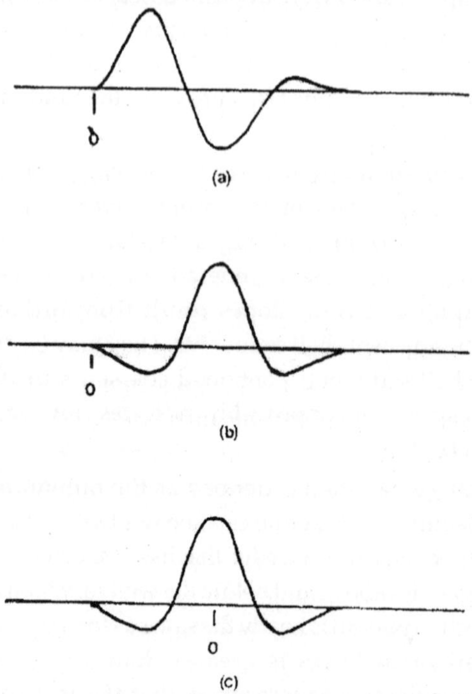

Fig. 13. Schematic diagram of minimum–phase and zero–phase wavelets (a) minimum–phase wavelet, (b) A symmetrical causal wavelet (c) zero-phase wavelet.

detectable limit, the minimum thickness for a layer to give a reflection, is of the order of 1/30 wavelength. Seismic frequencies are much lower than those used in well logging by a factor of about 100. Consequently, seismic resolving power is only one hundredth of that of well logging. In the shallow part of the earth, the velocity is generally small and the frequency is apt to be high, resulting in a wavelength of the order of 40 m, or resolvable limit of 10 m and detectable limit of 1.3 m. Deeper in the seismic section velocities are higher, e.g., 5 km/s, and frequencies lower, e.g., 20 Hz. Therefore, wavelengths are of the order of 250 m, with resolvable and detectable limits of 62.5 m and 8.3 m respectively. Thus, there is nearly a 10 to 1 ratio between wavelengths shallow and deep in the earth.

Seismic Attributes—Measurements of various aspects of the seismic data are called attributes. They are often displaced in seismic record section form useful hydrocarbon indicator attributes are amplitude, phase, frequency, polarity, and velocity. Complex trace analysis provides a method for calculating values for some attributes. The reflection strength display by colours shows the amplitude of the envelope of the seismic trace. The colours are arranged in a spectral sequence with the brightest (reds and oranges) indicating the largest values and the greens, blues, and whites indicating the smaller ones. The colours are related to numerical values in a quantitative way, each successive colour tone indicating one db of amplitude. Where the seismic data consist of separate primary reflections, the amplitude indicates the magnitude of the change in acoustic impedance at the reflecting interface. Often one of the problems with associating amplitude anomalies with gas accumulations is that small amounts of gas in a rock's pore space produce as large an acoustic impedance change as complete gas saturation. Hence amplitude bright spots may not indicate enough gas to provide a commercial accumulation. Phase is sometimes called instantaneous phase to avoid confusion with phase spectra, where phase is a function of frequency. A phase colour display emphasizes the continuity of data. Phase is independent of amplitude so that both weak and strong reflections show up equally. By emphasizing continuity or discontinuity, phase is especially useful in pinning down the locations of angular unconformities or faults. Phase or cosine of the phase is also often displayed in black and white form. A phase plot also forms the black background for colour plates.

Instantaneous frequency colour display is sensitive to the interference pattern between closely spaced reflections, and consequently instantaneous frequency often characterizes a particular sequence of reflectors. It is thus especially useful in character correlating from line-to-line or across faults. It is also sensitive to changes in the bedding sequence and thus is useful in telling where stratigraphic changes occur. The colour code for frequency uses the strong colours of red and orange to indicate low frequencies and the colours grade in a spectral sequence to blues indicating the higher frequencies, successive colour changes indicating 2Hz increments. Hydrocarbon accumulations often have exceptionally low frequencies immediately beneath them. This is referred to as a "low-frequency shadow". It is often a useful hydrocarbon indicator. The response on colour frequency displays is an orange or yellow shade. The distinction between frequency changes because of stratigraphic change and because of hydrocarbon accumulation is provided by the abruptness of the change, the former usually being distributed over a number of traces whereas the latter is usually abrupt. Faulting, of course, also produces abrupt change. An apparent-polarity colour display indicates the sign of the reflection coefficient if a reflection is an isolated primary reflection and if the wavelet is zero phase. The colour intensity is modulated by the reflection strength and thus shows the magnitude of the reflection coefficient also. The reflection from the top of the gas reservoir should be blue which indicates a negative reflection coefficient. Whereas that from the gas-water contact should be pink which indicates a positive reflection coefficient.

Seismic Data Processing—The recent prominence of seismic stratigraphy reflects our ability to obtain more information from seismic data. This is largely the result of cumulative improvements in seismic data acquisition and processing techniques. The major improvements have resulted from four things—(1) Common-Depth-Point (CDP) methods, (2) digital seismic data recording, (3) computer seismic data processing, and (4) data display improvements. There are four objectives of data processing which are given below:—

(a) The improvement of the signal-to-noise (S/N) ratio, either by enhancing the signal or attenuating the noise.

(b) The repositioning of data elements. The data on unmigrated seismic sections are referenced to the locations of the source and receivers, usually being plotted at the midpoint between them. Migration repositions data to the locations of the reflecting points.

(c) The measurements of attributes of the seismic data, which includes velocity, amplitude, frequency, polarity, and other measurements.

(d) The display of the seismic data as per the requirements of an interpreter, e.g., use of optimum scale for the particular interpretation objectives and displays, colour displays, etc.

Processing must be faithful to timing, amplitude and waveshape, because measurements of these determine the structure and give us stratigraphic information. Processing usually begins with an analysis of the seismic data. Analysis is used to understand the problems involved and to determine the processing parameters. Analysis is also done after processing to check that the processed records areas good as required by us. Processing parameters are :- (1) dereverberation to correct for multiple bounces in an overlying water layer, (2) statics correction to remove the effects of near surface variations, (3) normal move-out correction and common-depth-point (CDP) stacking based on velocity analyses, (4) application of two-dimensional (velocity) filtering to attenuate coherent noise wave-trains, and (5) coherency filtering to improve continuity. Obviously proper processing can considerably enhance our ability to draw correct stratigraphic conclusions. The interpreter should interact with the processor to make certain that objectives have been met in optimal fashion. While coherency filtering improves the continuity of events and is useful where following events is otherwise difficult, it obscures features such as faulting.

Quality control must include checks to ensure that processing decisions are consistent with interpretational objectives. One type of check is provided by a plot of the velocity that the computer applies in stacking data. Velocity analyses are usually run only every kilometre or so and then interpolated in both time and space by the computer. A display of the velocities that the computer applies allows the interpreter to check that they are reasonable from a geologic point of view. Linear inverse modeling is another display for control purposes. It shows selected velocity picks, the amount by which data will migrate, and the depth and the velocity of the interval above the event.

Seismic Facies Analysis—Facies are defined as the sum of all characteristics exhibited by a sedimentary rock and from which its origin and environment of formation may be inferred. The general aspect, nature or appearance of a sedimentary rock are produced under or affected by similar conditions, i.e., a distinctive group of characteristics that differs from other groups within stratigraphic unit. Seismic facies concerns those distinctive characteristics that can be seen in seismic data.

A seismic facies unit is a mappable three-dimensional unit of reflections whose characteristics differ from that of the adjacent facies. We use seismic facies distinction as well as unconformities to separate seismic sequence units. A seismic sequence unit is a package of reflections that results from the sediments within a time-stratigraphic depositional unit, i.e., they represent some certain age interval, whereas seismic facies describes the seismic characteristics that result from the depositional environment, e.g., marine or non-marine. The objective of seismic facies analysis is regional stratigraphic interpretation, especially, defining the depositional environment, the lithology, and the geologic history.

Sangree and Widmier provide a classification of the patterns for clastic facies. Table 1. This classification depends on the regional setting of units as well as their facies characteristics. Brown calls seismic facies interpretation a "process of elimination" in which the interpreter eliminates litho-facies/depositional environments based on inconsistencies with the area, basin venue, relationship to other units, etc, as well as reflection characteristics. In the interpretation process, special emphasis has to be given to lateral facies equivalents. An interpreter needs a good understanding of depositional processes, lithofacies composition, geometry, and spatial relationships. Brown uses internal reflection configuration as the first-order division.Table 2. Other items of seismic facies interpretation which should be included are (1) hemi-pelagic deposits are generally concordant at the base, (2) divergent reflections indicate either differential subsidence, faulting, or diapiric flow contemporaneous with

Table .1. Seismic Facies Analysis

Regional Setting	*Basic of Distinction*	*Subdivisions*	*Table 2*	*Interpretation*	*Other Characteristics*
Shelf	Reflection Character	High continuity high amplitude	1	Platfrom deposits– alternating meritic shales/ limestones	Possible cut by submarine canyones
	Unit Shape : widespread sheet or gentle wedge Reflections generally parallel/divergent			Interbedded hihg/low energy deposits Shallow marine clastics mainly by wave transport Fluvial clastics/marsh interbedded	Distinguish on basis of location compared to other facies
		Variable continuity low amplitude, occassional high amplitude	2	Fluvia/nearshore clastics, fluvial/wavetransport processes (delta platform)	Distinguish on basis of location compared to other facies
			4	Nonmarine clastics, fluvial/marginal marine transport	shale-prone if seaward of unit above sand-prone if seaward of unit below
		Low continuity, variable amplitude	3	Nonmarine clastics, fluvial/marginal marine transport	Occasional high amplitud and high continuity from coal members

Regional Setting	*Basic of Distinction*	*Subdivisions*	*Table 2*	*Interpretation*	*Other Characteristics*
	Mounded shape	Variable continuity and amplitude	10	Delta complex	Internal reflections gently sigmoid to divergent Occasional high amplitudes
		Local reflection viod	8	Ree	
Shelf margin prograded slope	Internal reflection pattern	Oblique fan shaped multiple fans	5	Adequate sediment supply Shelf margin deltaic in updip portions Occasionally due to stong currents in deep water	Moderate to high continuity and amplitude, reflections somewhat variable. Foreset (clinoform) dips to 10^0 (averaging $4\text{-}5^0$); steeper dips are calcareous multiple fans
		Sigmoid elongate lens/fan	6	Low sediment supply Low depositional energy	High continuity, high to moderate amplitude, uniform relfections
Basin slope, basin floor	Overll unit shape	Drape Sheet drape	7	Deep marine hemipelagic; mainly clays, low energy	High continuity, low amplitude Drapes over preexisting topography
		Mounded Contourite		Deep, Low energy	Variable continuity and amplitude

Regio-nal Setting	Basic of Distinction	Subdivisions	Table	Interpretation	Other Characteristics
		Fan–shaped		Variable energy, slimps/turbidity currents	Discontinous, variable amplitude At mouth of submarine canyons Composition depends on what was eorded up above
		Fill Slope front fill	11	Low energy, deep marine clays and silts	Variable continuity and amplitude Fan–shaped to extensive
		Onlapping fill		Low–velocity turbidity currents	High continuity variable amplitude
		Mounded onlap fill	13	High–energy turbidity currents	Discontinous, variable amplitude
		Chaotic fill		Variable–energy turbidites	Overall mound in a topographic low, gouge common in base Discontinuous, variable amplitude
		Canyon fill	12	Variable superimposed strata Coarse turbidites to hemipelagic	Variable continuity and amplitude Statigraphy often self –trapping

Table-2. Seismic Facies Analysis

Internal Reflection Configuration	*Possibilities*	*Table*	*Characteristics*
Parallel/divergent	Shelf platform alternating neritic shales/limestones	1	High continuity, high amplitude Uniform subsidence Upper surface may be eroded
	Delta platfrom sandstone/shale	2	Low/moderate continuity and amplitude except high amplitude for coals
	Alluvial plain/fan delta (meander belt, channel fill)	3	Discontinuous, variable amplitude
	Basin floor; hemipelagic shale/silt/clay	4	High continuity, low amplitude Drapes over preexisting topography
Progradational	High sediment supply; caly/silt/sand	5	Olique Variable continuity and amplitude On shelf or beyond shelf edge
	Low sediment supply; uniform conditions; calcareous clay/silt	6	Sigmoid High continuity, moderate uniform amplitude
Mounded/draped	Deep marine hemipelagic clay	7	Draped High continuity, low/moderate amplitude
	Biogenic carbone reefs	8	Shel/platfrom Elongate/subcircular shape Reflection free
	Canyon fan slump/turbidites	9	Slope/basin, fan–shaped Discontinous, variable amplitude

Internal Reflection Configuration	Possibilities	Table 1	Characteristics
Onlap/fill	Coastal onlap (paralic)delta/alluvial	10	Landward of shelf edge Low continuity, variable amplitude
	Marine onlap	11	Beyond shelf edge
	Canyon fill	12	In canyon
	Turbidite deposits	13	Basin floor

deposition, (3) formations underlying a shelf platform are generally sigmoidal whereas those underlying a delta platform are usually oblique, (4) delta platforms that extend into deep water usually involve growth faults, (5) prodelta sediments indicate water depths less than 100 m. In deep water, deposits are involved in density flow, or are otherwise reworked, and (6) canyon-fill deposits are usually turbiditie deposits interbedded with hemiplegics clays, so that they provide natural stratigraphic traps. The coarsest turbidities are usually near the base.

Seismic Log Manufacture—Inverse modeling is also based on the convolutional model. It assumes a noise-free seismic trace and attempts to drive the earth's reflectivity from it, then acoustic impedance values, and finally the geology. The process is called inverse modeling. This assumes that the amplitude of a seismic trace is proportional to the reflection coefficient and solves the reflectivity equation for the acoustic impedance, or by assuming the density, solves for the velocity. The frequency content of seismic data is severely band-limited compared to that of sonic logs. The result of inversion is usually called a seismic log or an instantaneous velocity log. The constraints applied are usually velocity so the result has the dimensions of velocity. The problems with trace inversion are mainly caused by the inability of separate completely reflectivity effects from noise. Trace inversion is mainly used to study trace-to-trace variations. Inversion assumes a noise-free input. The separation of wavelet and reflectivity remains a limitation to inversion. Inversion is essentially an operation on amplitude measurements, and amplitude depends on many factors besides reflectivit. Adequate handling of amplitude distorting factors is often the key to good inversion results. Migration is an important factor in remedying some types of amplitude distortion and thus properly precedes inversion.

Stratigraphic Patterns in Seismic Data—The nature of most rocks varies laterally, e.g., sand grades into sale, or lime content increases in a gradual facies change. As this occurs, we expect the rock properties to change and therefore, the details of a reflection from an interface bounding this rock member. Examining changes in individual reflections or small groups of reflections is called reflection character analysis. Ancillary measurements made on the data often emphasize character changes, and colour displays help localize such changes. Seismic sections can often be sub-divided into units that have common characteristics but that differ from adjacent units. The separation of seismic data into such units is the branch of seismic stratigraphy called seismic sequence analysis. The separate units are distinguished by different reflection characteristics, i.e., different seismic facies, the indicators of the depositional environment. The belt of oblique reflection patterns across the centre of fig.14, separates regular, fairly smooth reflection events above the belt from very irregular reflections below. The velocity of the sediments that constitute the oblique portion varies appreciably, and the variable travel time through this unit creates undulations and other irregularities in the deeper reflections. Higher velocity distinguishes the portions of the oblique reflection that are predominantly carbonate.

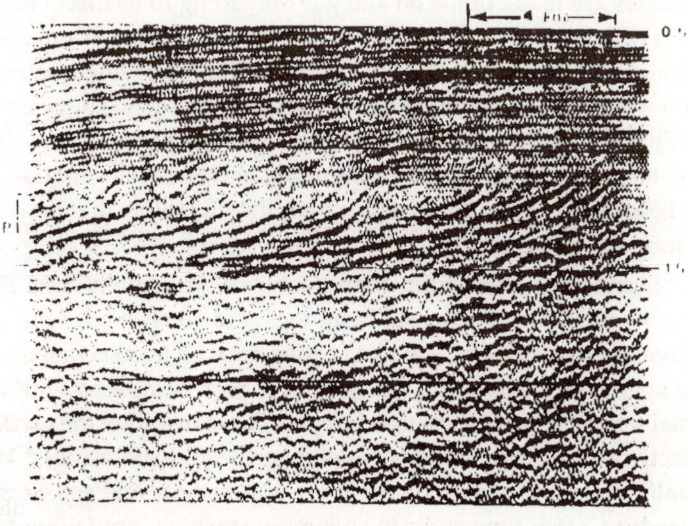

Fig.14. The belt of oblique reflection P has produced velocity anomalies in the deeper (After Sheriff, 1980).

Seismic sequence analysis alongwith seismic facies analysis and reflection character analysis represent the major techniques employed in seismic stratigraphy. Seismic sequence analysis isolates depositional units. Seismic facies analysis is generally converted with determining the depositional environment and lithotacies, concentrating on the up dip edges of units where depositional energy was probably greatest. Working out the geologic history is one of the key objectives of seismic stratigraphy. Reflection character analysis concentrates on changes involving single or a few reflections and endeavour to interpret these in terms of stratigraphic changes or hydrocarbon accumulations. It thus is involved in predicting reservoirs and traps.

Improved seismic data quality has made stratigraphic interpretation from seismic observations possible. However, seismic interpretation is subject to a number of important limitation. Of special importance are those involving five things like things (1) resolution both vertical and horizontal, (2) wavelength or land pass, (3) noise, (4) out-of-the-plane reflections, and (5) velocity variations. When estimates are made of the oil and gas remaining to be discovered, that which is trapped stratigraphically almost always dominates, although different individuals also estimate widely verying amounts. In the record from a borehole, some portions differs to suggest the possibility of a stratigraphic trap nearly. Reflection character analysis can help us locate the nearby accumulation that a well suggests. It can help us search for stratigraphic traps directly rather than relying on luck and statistics.

Synthetic Seismogram Manufacture—We usually take the earth's reflectivity as derived from logs and convolve it with an equivalent wavelet. The wavelet is often derived from actual data. The synthetic seismogram that results is then compared with an actual seismic record. If the match is poor, it means either the earth's reflectivity or the equivalent wavelet shape needs to be changed. We usually make the changes we feel are appropriate and repeat the procedure, often repeatedly in an iterative fashion, until we judge the degree of match acceptable. Agreement between the actual and synthetic seismograms may not be perfect because the actual and synthetic differ in various regards, such as the effects of density variations, noise contributions, and the fact that they may not be looking at exactly the same sub-surface volumes. When a good match can be made, the geologic model can then be varied to see what effects changes in the geology would produce on seismic records. This is done to help identify observed changes on the record with possible changes in the stratigraphic sequence.

Conceptually the procedure is simple and straightforward, but actually the various interfaces that contribute are so close together, and their affects overlap so much that it is usually difficult to associate a seismic event with any particular interface. Fig. 15. The problems in actual synthetic seismogram manufacture mainly result from incomplete and inaccurate data. Density logs are usually not available. Density is often inferred from the velocity. Usually only velocity or sonic changes are taken into consideration and usually

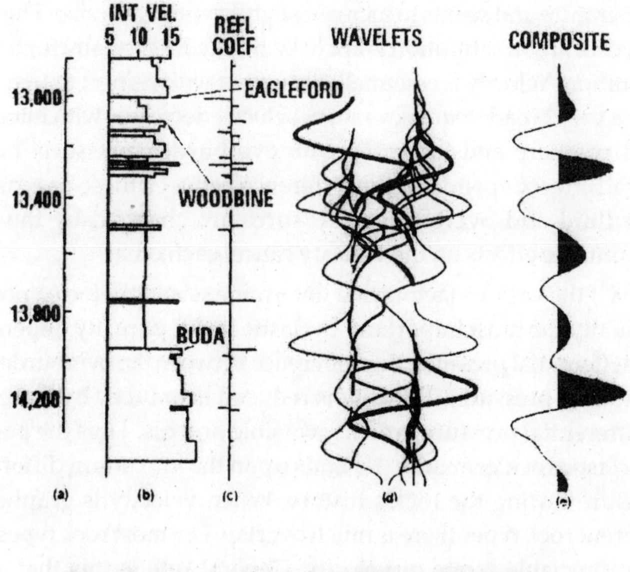

Fig. 15. Construction of synthetic seismogram (After Vail, Todd and Sangree, 1977).

the sonic logs are not complete because the shallow parts of the borehole were not logged. The penetration of sonic logs into the formation is not always adequate to represent the unaltered formations reliably. Because disagreements between sonic logs involving different spacings or penetration are seen. Reflections involve a large portion of the subsurface and the borehole may not adequately represent the subsurface which affects the seismic trace. Multiples and other types of noise are different on synthetic and actual data. Despite all their limitations, synthetic seismograms are important tools in stratigarphic interpretation and hydrocarbon location.

Velocity Measurement—Velocity measurements are often used to determine the nature of sediments, so it is appropriate to examine the sensitivity of velocity to those factors we wish to determine and also to other factors that affect the reliability of determinations. The decrease in velocity with increasing porosity is usually marked. While velocity increases with density, the range of density values is small. Velocity decreases slightly with increase in temperature and seems to increase slightly with grain size. The effect of hydrocarbon saturation is apt to be marked, especially for low gas saturation. Velocity is essentially invariant with respect to frequency over a very broad frequency range. Velocity decreases with interstitial fluid pressure and increases with overburden pressure, but the overall effect depends on the difference between these pressures. If pore fluid and overburden pressures are changed by the same amount, the effects on the velocity cancel each other.

Of the various factors that determine seismic velocity, porosity is usually the most important. In clastic rocks, porosity depends on the defferential pressure, the defference between the overburden and interstitial pressures. Porosity is reduced is reduced by increasing of differential pressure in an irreversible process. Thus the porosity of a clastic rock generally depends upon the maximum differential pressure during the rock's history. When velocity is graphed for different rock types there is much overlap, i.e., most rock types have an appreciable range of velocity. General rule is this that clastic rocks have low velocities and carbonates or evaporates have high velocities. Velocity alone is generally not sufficient to distinguish rock type. High porosity values generally are associated with low velocity and low porosity values are generally associated with high velocity. The porosity of a clastic rock generally decreases with depth of burial, but subsequent uplift does not change the porosity or the velocity significantly. Porosity generally becomes smaller as rock sorting becomes poorer and as cementation increases. The spontaneous potential (SP) is used to distinguish between sand and shale units. A best fit curve drawn through the velocity of the shale units indicate lower velocity than a best-fit curve drawn through the velocity of the sand units. However, within each rock type, the variation is greater than the difference between rock types. Thus it may be possible to make a statistical analysis of sand shale distribution while, for any individual sample, prediction based on velocity measurement is very uncertain. The nature of the interstitial fluid affects velocity. Gas has more effect than oil in lowering velocity. Only a small amount of gas in the pore space produces a marked effect on velocity, but further increase the gas saturation has only a minor effect. Thus very small amounts of gas can produce large reflectivity and, consequently, amplitude anomalies nearly as strong as those associated with commercial reservoirs. Porosity is the most important stratigraphic feature that influences velocity. Shale content tends to lower velocity, lime and dolomite content to increase velocity. Thus a shaley limestone is apt to have a lower velocity than a sandy limestone, and a limey sand had higher velocity than a lime-free sand.

Velocity measurements are usually based on the change in the arrival time of reflections as the shot-to-geophone distance changes. This distance is called offset, the time difference because of offset is called normal move-out (NMO). The velocity that the normal move out implies is called stacking velocity because it is what yields the optimum common depth point (CDP) stack. The calculation procedure is called velocity analysis. Velocity analysis is a fairly expensive process, and hence an analysis is often run only every 1 to 3 kilometres and the intervening velocities are interpolated. Uncertainties in individual analysis sometimes produce fictious velocity anomalies. To reduce the effects of noise, data over a small region are often averaged in making an analysis. This results in less scatter but then the measurements are averages that do not apply at specific locations. Velocity can also be calculated from amplitude information by trace inversion.

Wavelet Processing—Wavelet processing is applied usually with one of three different kinds of objectives : (1) Equalizing the wavelet, i.e., making the wavelet shape the same everywhere, (2) Replacing the effectire wavelet with some other wavelet, usually a zero-phase one, and (3) Separating the reflectivity of the earth from the seismic wavelet. Wavelet processing ordinarily begins with determining the effective wavelet shape, which is usually the critical and difficult aspect of wavelet processing. After an operator is determined to change the effective wavelet to some desired wavelet. The desired wavelet is usually a very short one, hence the processing is sometimes called wavelet compression. The width in time and the width in frequency have a reciprocal relationship. Usually a bandwidth less than 1.5 octaves is inadequate and a bandwidth of 2.5 octaves or more is desirable.

Wavelet processing in order to equalize the effective wavelet is often done in a prestack mode. The objective is to make the effective wavelet shape the same for all the traces that are to be stacked together. If the waveshapes differ, the low-frequency components tend to be stacked in phase but the high frequency components are apt to be stacked out of phase and therefore attenuated. The effect of waveshape variations in stacking is that of a high frequency filter that narrows the bandwidth and limits the resolution. Sometimes the source waveform is recorded in the field, in which case a deterministic deconvolution can be used. With land data, the waveform has to be determined by statistical analysis of the data. Most statistical analysis procedures assume minimum phase. While most impulsive sources, are nearly minimum phase, a few non-minimum-phase roots are usually involved. Better results can be achieved if minimum phase does not have to be assumed. The removal of the effects of changes in source waveshape are important in distinguishing waveshape variations produced by stratigraphic features.

Wavelet processing with the object of separating the earth's reflectivity from seismic wavelet effects is based on the convolutional model. The convolutional model assumes that a seismic trace is simply the result of convolving the earth's reflectivity with an effective wavelet plus additive-noise. This type of wavelet processing implies that systematic noise has been completely removed and so usually follows the various noise attenuating processes, such as statics correction, stacking, deconvolution, and so on, as usually one of the last processes before display. Three types of products result from this post-stack wavelet processing: –

1. A high-resolution zero-phase section,
2. The earth's reflectivity,
3. The effective wavelet shape.

Each of these may be processed further. The reflectivity may be input to trace inversion in order to stimulate acoustic impedance logs, and analyzed for trace-to-trace variations attributable to stratigraphic changes. The derived wavelet shape may be combined with sonic logs in the manufacture of synthetic seismograms for reflection character prediction studies. Other processes are deterministic processing to remove instrumental filtering effects and source-wave- shape corrections that assume that the source was everywhere constant.

References

1. Becquey M., Lavergne M., and William. C., 1979; *Acoustic impedance logs computed from seismic traces*; Geophysics, Vol. 44, p. 1485-1510.
2. Brown L.F., and Fisher W.L., 1979; Principles of seismic stratigrpahic interpretation ; Notes for AAPG Petroleum Exploration School.
3. Brown L.F., and Fisher W.L., 1977; Seismic stratigraphic interpretation of depositional systems : examples from Brazilian rift and pull apart basins; AAPG Memoir 26, pp-213-245.
4. Bubb J.N., and Hatlalid W.G., 1977; Seismic recognition of carbonate buildups; AAPG Memoir 26, pp. 185-204; also in AAPG Bulletin, Vol. 62, p. 772-791.

5. Dorbin M.B., 1977; Seismic exploration for stratigraphic traps; AAPG Memoir 26, pp. 329-352.
6. Gardner G.H.F., Gardner L.W., and Gregory A.R., 1974; Formation velocity and density—the diagnostic basis of stratigraphic traps; Geophysics, Vol. 39, pp. 770-780.
7. Lindseth R.O., 1979; Synthetic sonic logs—a process for stratigraphic interpretation; Geophysics, Vol. 44, pp.3-26.
8. Lyons P.L., and Dobrin M.B., 1972; Seismic exploration for stratigraphic traps; AAPG Memoir 16, pp. 225-243.
9. Marr J.D., 1971; Seismic stratigraphic exploration; Geophysics, Vol. 36, pp. 311-329, 533-553, 676-689.
10. Meckel L.D., and Nath A.K., 1977; Geologic considerations for stratigraphic modeling and interpretation; AAPG Memoir 26, pp. 417-438.
11. Mitchum R.M., and Vail P.R., 1977; Seismic stratigraphic interpretation procedure; AAPG Memoir 26, pp-135-144.
12. Mictchum R.M. , Vail P.R., and Thompson S., 1977; The depositional sequence as a basic unit for stratigraphic analysis; AAPG Memoir 26, pp. 53-62.
13. Ramsayer G. R., 1979; Seismic stratigraphy, a fundamental exploration tool; Offshore Technology Conference paper 3568.
14. Sangree J.B., and Widmier J.M. , 1977; Seismic interpretation of clastic depositional facies; AAPG Memoir 26, pp.165-184.
15. Sangree J. B., and Widmier J. M., 1979; Interpretation of depositional facies from seismic data; Geophysics, Vol. 44, pp. 131-160.
16. Schramm M. W., Dedman E. V., and Lindsey J. P., 1977; Practical stratigraphic modeling and interpretation; AAPG Memoir 26, pp. 477-502.
17. Sheriff R. E., 1974; Seismic detection of hydrocarbons—the underlying physical principles; Offshore Technology Conference paper 2001.
18. Sheriff R. E., 1975; Factors affecting seismic amplitudes; Geophysical Prospecting, Vol. 23, pp. 125-138.
19. Sheriff R.E., 1977 a; Limitations on resolution of seismic reflections and geologic detail derivable from them; AAPG Memoir, 26 pp. 3-14.

20. Sheriff R. E., Taner M. T., and Rao K., 1978; Seismic attribute measurements in offshore production; Offshore Technology Conference paper 3174.

21. Sheriff R.E., and Taner M.T., 1979; Seismic attribute measurements help define reservoirs; Oil and Gas Journal, Aug. 20, 1979; pp. 182-186.

22. Sheriff R.E., 1980; Nomogram for Fresnel zone calculation; Geophysics, Vol. 45, pp. 968-992.

23. Sheriff R.E., 1980; Seismic stratigraphy; International Human Resources Development Corporation, Boston, p. 227.

24. Taner M. T., and Sheriff R. E., 1977; Application of amplitude, frequency and other attributes to stratigraphic and hydrocarbon determination; AAPG Memoir 26, pp. 301-328.

25. Taner M. T., Kochler R., and Sherriff R. E., 1979; Complex seismic trace analysis; Geophysics, V.44, p. 1041-1063.

26. Vail P. R., Mitchum R. M., and Thompson S., 1977 a; Relative changes of sealevel from coastal onlap; AAPG Memoir 26, pp. 63-82.

27. Vail P. R., Mitchun R. M., Thompson S., 1977 b; Global cycles of relative changes of sea level; AAPG Memoir 26, pp. 83-98.

28. Vail P. R., Todd R. G., and Sangree J. B., 1977; Chronostratigraphic significance of seismic reflections; AAPG Memoir 26, pp. 99-116.

Annexure : A

Technical Terms Used in Seismic Stratigraphy

A

Acoustic Impdance—The product of seismic velocity and density.

Aggradation— Building upward by deposition.

Anti-alias Filter—A filter used before sampling to remove high frequencies that otherwise would cause ambiguities in information content. "Alias" and "ant alias" filter mean the same thing.

Apparent Velocity—The velocity that a waveform appears to have in some direction, usually in the horizontal direction and usually not the direction of the wave.

Attributes (Seismic)—Measurements of properties of seismic data, such as amplitude, frequency, velocity, polarity, and so on.

B

Back-reef facies—Deposits on the landward side of a reef.

Baselap—Angular termination of strata at the lower boundary of a depositional sequence; includes both onlap and downlap.

Biogenic Rock—Rock manufactured directly by organisms.

Bioherm—A mound-like or reef-like mass of rock built up by sedentary organisms.

Bright Spot—(1) A local build-up of amplitude, which sometimes indicates a hydrocarbon accumulation; (2) The class of hydrocarbon indicators, one of which is an amplitude anomaly.

C

Carbonte-build-up Configuration—A seismic reflection configuration, usually a mound. Interpreted as a carbonate reef or bank.

Chaotic Configuration—Discontinuous, discordant seismic reflection pattern interpreted either as strata deposited in a highly variable, high energy setting, or as strata that have been so deformed that continuity has been disrupted.

Chronographic/Chronostratigraphic Chart—A summary chart on which geologic time is plotted vertically and distance horizontally and on which stratigrpahic information is brought together.

Clastics— Sediments composed of fragments transported from their point of origin.

Clinoform Surface—A sloping depositional surface, commonly associated with strata prograding into deep water.

Coastal Deposits—Sediments deposited near sea level (littoral, paralic, or coastal no marine).

Coastal Onlap— The progressive landward onlap of coastal (littoral or coastal no marine) deposits. Coastal aggradation and coastal encroachment are the vertical and horizontal components of coastal onlap, respectively.

Coherent— Having a fixed phase relationship to each other.

Coherence Filter—A multichannel filter that enhances coherent events.

Common-depth Point Stacking— The combination of data that have in common the same midpoint between source and detector.

Competent Rock—A rock that maintains its shape compared to adjacent rocks when subjected to a tectonic force.

Compressional Wave—A seismic wave in which particle motions is in the direction of the wave.

Concordance—Parallelism of strata to sequence boundaries.

Constructive Shelf —A shelf where sediment influx exceeds marine erosional effects.

Convolution—A mathematical process constituting linear filtering. Natural processes in the earth appear to accomplish convolution and many instrumental and processing procedures involve convolution. Linear superposition constitutes convolution.

Convolution Model—The concept that a seismic trace can be represented by the convolution of an equivalent wavelet and the earth's reflectivity, with noise added.

Cycle of Relative Change of Sea Level—The interval of time during which a relative rise and fall of sea level takes place. Three orders of cycles are recognized : first-order, supercycle, and cycle.

D

Disconvolution—Removing the effect of an earlier filtering action. "Predictive" disconvolution involves the prediction and removal of multiples that involve the near surface as one of the reflectors. "Spiking" disconvolution attempts to achieve an equal amplitude spectrum at all frequencies and sharpen the effective wavelet. Wavelet processing can be thought of as a type of disconvolution.

Destructive Shelf —A shelf where the effects of marine erosion exceed those of sediment influx.

Detectable Limit—The minimum thickness for a bed to give a seismic reflection, in the vicinity of 1/30 of the dominant wavelength.

Differential Compaction—Uneven settling of sediments as a result of reduction of porosity. Sands/shales generally shrink more under pressure than reef limestones.

Diffraction—The energy returned from a point, such as the abrupt termination of a reflector at a fault, unconformity, reef, salt dome, and so on. Diffraction energy does not satisfy Snell's Law at the diffracting point.

Dim Spot—A local lowering of amplitude, sometimes because of hydrocarbons being locally present in a rock where the acoustic impedance is appreciably greater than that of the overlying rock.

Direct Hydrocarbon Indicator—Something in seismic data that may indicate an accumulation of gas or oil.

Divergence Correction—Correction applied to amplitude values to compensate for the decrease in energy density with distance from the source, that is to compensate for geometrical spreading.

Divergent Reflections—Seismic reflections characterized by a laterally thickening wedge-shaped unit in which thickening is accomplished by splitting of individual reflection cycles within the unit, rather than thickening predominantly by onlap at the base or toplap or erosion at the top of the unit.

Downlap— A relation in which initially inclined strata terminate down dip.

Downward Shift of Coastal Onlap—A seaward shift from the highest position of coastal on lap in a seismic sequence to the first onlap in the next younger unit. It is used to recognize relative falls of sea level.

Drape—A unit draped evenly over a preexisting surface without regard for the preexisting topography.

E

Earth Reflectivity —The time series that is the sum of the reflection coefficients at all interfaces spaced according to reflection times. Also called a "Stickogram".

Eustatic Cycle—The interval of time during which a worldwide rise and fall of sea level takes place.

Eustatic Level Chart– A chart showing worldwide changes in sea level.

F

Facies – The sum of all characteristics exhibited by a sedimentary rock and from which its origin and environment of formation may be inferred; the general aspect, nature, or appearance of a sedimentary rock produced under or affected by similar conditions; a distinctive group of characteristics that differs from other groups within a stratigraphic unit.

Fill Configuration – Seismic reflections interpreted as strata filling a negative relief feature. Fill units are classified by external form (channel fill, trough fill, basin fill, slope-front fill; or by internal reflection configurations (onlap, mounded onlap, divergent, prograding, chaotic, complex).

First Order Cycle– A cycle of relative or eustatic change in sealevel that has a duration in the order of 200 to 400 million years.

Fluvial–Relating to a river, as with deposits laid down by a stream.

Fondo form–The seaward portion of a progradational pattern; foreset.

Fore-reef Facies–Sediments laid down seaward of a reef.

Forward Modelling–Computing the effects of a model, as in synthetic seismogram manufacture.

Frequency Spectrum – The amplitude and phase of sine waves of different frequencies that would add up to a particular waveform.

Fresnel Zone–The portion of a reflector responsible for a reflection event seen at a point. Reflection ray paths from source to detector that differ by less than 1/2 wavelength interfere constructively; the portion of the reflector from which they can be reflected constitutes a "Fresnel zone".

G

Gardner's Relation – The relationship that density is proportional to the 1/4 or quarter power of seismic velocity.

H

Hemipelagic Facies – Deep-sea sediments, mostly very fine grained.

Hiatus – An interval of geologic time not represented by strata at a specified position. If the hiatus encompasses a measurable interval of geologic time, the surface is an "unconformity". A "nondepositional" hiatus refers to an interval during which no strata were deposited, an "erosional" hiatus to an interval during which strata were removed by erosion.

Hinterland Sequence – A sequence of non-marine deposits laid down at a site interior to the coastal area, where depositional mechanisms were not controlled by sea level.

Hummocky Clinoform – A reflection configuration that consists of irregular, discontinuous, subparallel reflections forming a practically random, hummocky pattern. This pattern commonly grades laterally into larger, better-defined clinoform patterns, and upward into parallel reflections.

Hz – Full form is hertz or cycles per second.

I

Impulsive Source – A source of very short duration.

Interference – The superposition of waveforms is "constructive" where they add in-phase and "destructive" where they add out-of-phase.

Inverse Modelling – Calculating a model from observed effects.

Inversion – Conversion of reflectivity to acoustic-impedance values.

Instantaneous Frequency – The time rate of change of phase. Instantaneous frequency is a tool for seeing lateral stratigraphic or hydrocarbon variations.

Isopach—A line drawn through points of equal thickness for a unit.

Isotime—(1) Sometimes used for isopach where thickness is measured in reflection time difference. (2) Isotime surface is sometimes used for a surface of constant time.

L

Linear Inverse Model – A type of migrated section.

Littoral—Relating to the depth zone between high and low water.

Lowstand—A time when sea level is below the shelf edge.

M

Migrating Wave—A reflection configuration consisting of superimposed wave-shaped reflections, each progressively offset laterally from the preceding.

Migration—Repositioning of reflected energy so that it indicates the location of the reflecting point; also called "imaging".

Minimum Phase—A characteristic of waveforms that have their energy concentrated early in the waveform. A minimum phase wavelet has energy build-up at front of wavelet.

Model—A concept of a relationship between cause and effects.

Mound Configuration—A reflection configuration interpreted as strata forming an elevation above the general level of surrounding strata.

Multiples—Energy that is reflected more than once.

N

Neritic—Relating to water depths between low tide and 200 m.

Noise—Any unwanted energy or signal.

Normal Incidence—The situation where the direction of travel of a wave is perpendicular to an interface.

Normal Moveout (NMO)—Differences in the arrival time of reflections because of the distance between source and receiver.

O

Oblique Reflection—A reflection configuration interpreted as a prograding (clinoform) pattern whose upper surface is near the wave base. Strata terminate updip by toplap and downdip by downlap. Successively younger foreset segments build laterally from a relatively constant upper surface characterized by lack of top-set strata. The upper part of oblique sequences are apt to be sand prone.

Offlap—A pattern in which each successively younger unit leaves exposed a portion of the older unit on which it lies interpreted as strata prograding into deep water.

Offset—The distance between source and detector.

Onlap—A base-discordant relationship in which initially horizontal or inclined strata terminate updip against a surface of greater initial inclination. "Proximal" onlap is onlap in the direction of the source of clastic supply; "distal" onlap is in the direction away from the source; "marine" onlap is onlap of marine strata; "coastal" onlap is the progressive landward onlap of coastal deposits.

Out-of-Phase—A relation between two waveforms where the phase differs, usually by nearly 180 degree.

P

Paracycle—The interval of time occupied by a rise and still-stand of sealevel, followed by another rise, with no intervening fall.

Parallel Configuration—Parallel seismic reflections within a seismic sequence.

Pelagic Deposits—Sediments containing little of continental source, being mainly from material in general oceanic suspension.

Phase—(1) The amount of rotation in circular motion, or the stage in the cycle of a harmonic or pseudoharmonic waveform. Phase carries the timing information in seismic data (2) An "instantaneous phase" plot shows phase (often in colour-coded form) or some function of the phase (such as the cosine of the phase) as a function of arrival time. A phase plot emphasizes the continuity of coherent events or abrupt interruptions of continuity.

Polarity—The sign of the reflectivity (positive or negative) that indicate whether the acoustic impedance is increasing or decreasing at an inteface. Apparent polarity is the polarity if a single interface is involved and if the effective wavelet is zero phase.

Predictive Disconvolution—A process that uses early arrivals to predict and remove later multiples involving the same reflectors.

Prodelta—The seaward part in front of a delta complex, lying below the wavebase.

Progradational Configuration—Seismic reflections interpreted as deposition due to lateral outbuilding (and sometimes upbuilding). Progradational **reflection** patterns include sigmoid, oblique, complex and so on.

R

Reef—A ridge-like or mound-like structure built by sedentary organisms.

Reflection Character Analysis—Determining stratigraphy or hydrocarbon accumulations from the detailed seismic waveshape.

Reflection Configuration—The geometric pattern of seismic reflections, interpreted to represent configuration of strata.

Reflection-free—Absence of reflections, interpreted as due to homogeneous, nonstratified, highly contorted, or very steeply dipping geologic units.

Reflectivity—The ratio of the amplitude of a reflected wave to that of the incident wave. For normal incidence (direction of wave perpendicular to interface), the reflectivity is given by the ratio of the change in acoustic impedance to twice the average acoustic impedance. Also called "reflection coefficient". Reflectivity is generalised to include the effects of a series of interfaces.

Regression – A seaward movement of the shoreline indicated by the littoral facies moving seaward with time.

Relative Change in Sealevel – An apparent rise or fall of sea level with respect to the land surface. Either sealevel rises or the land surface subsides, or both.

Relative Fall of Sealevel – An apparent fall of sealevel with respect to the surface of deposition. Recognized by a downward shift of coastal onlap.

Relative Rise of Sealevel – An apparent rise of sealevel resulting from sealevel rising or the land subsiding. Recognized by coastal onlap.

Relative Stillstand of Sealevel – An apparent constant stand of sealevel with respect to the surface of deposition. Recognized by coastal toplap and lack of coastal onlap.

Resolvable Limit – The minimum thickness of a bed for the effects of the bed's upper and lower surfaces to give distinguishable reflection effects, considered as 1/4 or quarter of the dominant wavelength.

Resolution – The limit at which the effects of two features can be distinguished from the effects of one feature.

Reverberation – Energy that bounces back and forth within a layer, especially a water layer.

Ricker Wavelet – A particular zero-phase wavelet of convenient mathematical formulation, which is often used in synthetic seismogram studies.

Ringing Character – A wavelet that has more cycles than usual.

S

Stand-prone Seismic Facies – Seismic facies interpreted as deposited in a clastic environment with sufficiently high energy to sort sand (assuming sufficient quantities of sand-sized particles were available).

Seismic Facies – Distinctive features characterize seismic reflection, such as continuity, amplitude, configuration, and so on.

Seismic Facies Analysis – The description and geologic interpretation of seismic-reflection parameters, including configuration, continuity, amplitude, frequency, and interval velocity, and interpretation as to the implied depositional environment.

Seismic-Facies Map – A map that shows the areal distribution, configuration, thickness or other aspect of a given seismic-facies unit or parameter.

Seismic-Facies Unit – A mappable three-dimensional unit of reflections whose characteristics differ from those of adjacent facies. Sometimes used interchangeably with "seismic sequence".

Seismic Log – Acoustic impedance or velocity calculated from observed seismic data by inversion, making certain assumptions. Also called "SAILE trace", "synthetic acoustic impedance log or synthetic sonic log, or "instantaneous velocity".

Seismic Sequence – A depositional sequence identified on a seismic section by mapping the bounding unconformities. Sometimes used interchangeably with "seismic facies unit".

Seismic Sequence Analysis – The identification of deposit packages of concordant reflection separated by surfaces of discontinuity, and interpreting them as depositional sequences.

Seismic Stratigraphy – The study of stratigraphy and depositional facies as interpreted from seismic data.

Sequence – A term applied to a relatively conformable succession of genetically related strata bounded at top and base by unconformities or correlative conformities. A "depositional" sequence (time-stratigraphic unit) is a stratigraphic unit. A "seismic" sequence is a depositional sequence identified on a seismic section.

Shale-Prone Seismic Facies– Seismic facides interpreted as clastics deposited where depositional energies are insufficient to produce significant sorting.

Shear Wave– A seismic body wave in which particles move at right angles to the direction of the wave.

Sheet – A thin, widespread, tabular unit

Shingled Configuration–A prograding seismic pattern with parallel upper and lower boundaries and very gently dipping, parallel, oblique, internal reflections that terminate by apparent toplap and downlap.

Sigmoid Configuration–A prograding clinoform pattern formed by superposed sigmoid (S-shaped) reflections, interpreted as strata deposited in deep, quiet water. Usually shale prone.

Signature Correction–A process to change observed seismic data into what would have been obtained with a desired wavelet shape, used where the wavelet shape was known.

Singing Record–A record that involves reverberation effects.

Sonic Log–A record of the transit time over an interval.

Stacking–Combining the data from different records.

Stacking Velocity–The velocity that produces the optimum stack for certain events. Stacking velocity is determined in velocity analysis. For horizontal velocity layering, the stacking velocity is a root-mean-square average of the velocities. Stacking velocity is often called erroneously "rms velocity" where the foregoing assumption is not satisfied.

Static Correction—A correction for variations in arrival time because of near-surface thickness, velocity, elevation, or datum variations.

Stratal Configuration—The geometric for patterns and relations of strata within a stratigraphic unit, indicative of depositional environment (and later structural movement).

Stratal Surfaces—Surfaces that separate sedimentary strata. They may represent short periods of non-deposition or a change in the depositional regime.

Stratigraphy—It is the branch of geology that deals with the definition and description of major and minor natural devisions of rocks, and also with the interpretation of their significance in geologic history. It deals specifically the geologic study of the form, arrangement, geographic distribution, chronologic succession, classification, and especially correlation and mutual relation ship of rock strata. Also it deals in terms of their origin, occurrence, environment, thickness, lithology, composition, fossil content, age, history, paleogeographic condition, relation to organic evolution, and relation to other geologic concepts.

Stratum—A tabular layer of sedimentary material visually separable from adjacent layers by a change in character. A general term that includes both "bed" and "lamination". The term is more frequently used in its plural form, "Strata".

Supercycle—A group of cycles of relative change in sealevel in which a cumulative rise to a higher position of sealevel is followed by a cumulative fall to a lower position. The duration of a supercycle is 10 to 30 million years. A "second order cycle".

Super-position — The principle that the result of an operation is the same as the sum of the results of the operation carried out in pieces.

Surface Consistent—Assignment of time delays or amplitude attenuation to source or detector locations.

Synthetic Seismogram – A seismic trace or section calculated from a model.

T

Third Order Cycle – A cycle of change of sealevel that has a duration of 1 to 10 million years, the fundamental cycle that represents only one rise and fall; a "cycle".

Time-Average Equation – An empirical relation for the velocity in a two component system, the two components usually being the matrix and the fluid in the matrix's pore space. If V_m is the velocity of the matrix material, V_f the velocity of the fluid and ϕ the porosity, then the velocity V is given by the following equation –

$$V = \frac{(1-\phi)}{V_m} + \frac{\phi}{V_f} \qquad \text{................(1)}$$

Time-Stratigraphic Unit– A three-dimensional set of facies deposited contemporaneously as parts of the same system, genetically linked by depositional processes and environments. Also called a "depositional sequence".

Time Surface – A surface that at one time was the surface of the solid earth. Strata and seismic reflections have the attitude of time surfaces. Also isochronous and stratal surface.

Toplap—Termination of strata against an overlying surface mainly as a rasult of nondeposition (Sedimentary bypassing) with only minor erosion.

Two-Dimensional Filtering—The attenuation of wavetrains based on their apparent velocity or apparent dip; a type of coherence filtering.

Transgression—A landward movement of the shoreline indicated by landward migration of littoral facies.

Transit Time—The reciprocal of seismic velocity.

Truncation–Termination of strata or seismic reflections along an unconformity surface due to postdepositional erosion.

Turbidite–A sediment deposited by a turbidity current.

Turbidity Current–A density current of water with other matter in suspension that sometimes flows swiftly down sub-marine slopes.

U

Unconformity–A surface of erosion or nondeposition that separates younger strata from older rocks and represents a significant hiatus (at least a correlatable part of a unit) not represented by strata. Unconformities occur at each global fall of sea-level, although in some areas of continuous deposition, the hiatus may be too small to detect seismically.

Undaform – The subaqueous landform produced by wave action.

V

Velocity– The speed with which seismic waves travel.

Velocity Analysis – A procedure for determining the stacking velocity (or normal moveout) that will optimize certain events on a stacked section.

Velocity Anomaly – A distortion of seismic data (especially of arrival times) because of changes in velocity in the horizontal direction.

Velocity Filtering–Attenuating events on the basis of dip moveout.

W

Wave Equation Migration – A processing method to migrate seismic data by downward continuation of the wave-field by numeric solution of the wave equation.

Wavelet Compression – Processing with the objective of producing the seismic section that would result if the wavelet had been very short (ideally, impulsive).

Wavelet Processing – Processing that involves determining the seismic wavelet shape and/or changing the effective wavelet shape. Wavelet processing can mean any of a number of different processes with different objectives.

Wiener Filtering – Processing so as to change a waveform into some desired waveform as nearly as possible in a least-squares sense.

Z

Zero Phase – A characteristic of waveforms that are symetrical. It is a symmetrical wavelet of same length and frequency content as minimum phase wavelet but the time scale is shifted.

Four

Log Analysis of Subsurface Geology

Introduction

The first log was run to resolve a specific geological problem. The experiment succeeded in showing distinctive relationships between borehole resistivity measurements and lithology which were useful for stratigraphic correlation. Since that time, the major emphasis of log interpretation has shifted to the location and evaluation of porous, permeable hydrocarbon zones and a reservoir engineering perspective. Now there has been a growth of interest in the broader implications of logs as aids in the exploration for stratigraphic traps. Subtle changes in depositional facies and diagenetic patterns are the key to many modern plays. Logs are physical measurements of rocks.

The range of logging devices has expanded dramatically since the early days to include the measurement of a host of electrical, thermal, nuclear, and acoustic properties. Consequently, the logging program of a typical modern well records a wealth of numerical information from subsurface formations. The results are presented an analog traces of variation, drawn as functions of depth with scaling conventions which were observed by all logging service companies. These conventions have been chosen both to maintain a common standard and to allow the simultaneous presentation of a number of log traces in a meaningful manner. The multiple presentation format is a useful device in the determination of trends, patterns, and discriminations. The responses of logs to a variety of idealized lithologies are themes that are the basis of repeated variations along the record of a stratigraphic sequence. The major composite variations in fluctuation of the log suite carry the "melody" of the interbedded lithologies and their gross reservoir characteristics. Differences between the logs are "harmonic" elements which convey useful information concerning changes in mineralogy and fabric. The use of a computer for log analysis is the obvious means to condense the large information base to transformations which are meaningful to both geologists and reservoir engineers. Useful results require a skillful integration of geology, mathematics, and log analysis.

Log traces are used in the delineation of boundaries of subsurface units for the production of structure and isopach maps. The "shapes" of log profiles within units (particularly the SP, gamma ray, and resistivity) are qualitative aids in the interpretation of depositional origin, mostly in sandstone-shale sequences. Most observations of sedimentary rocks are purely descriptive, with assignations to broad categories of mineral composition and texture. The measurements of logging tools are in immediate numerical form and so are suitable material for data analysis. The process of logging a hole can be considered as remote sensing of the subsurface. The depth axis of each log is some function of geological time. The treatment of logging data by conventional time-series methods is complicated by the vagaries of depositional histories as a result of varying rates of sedimentation and periods of erosion and nondeposition. Reasonable geological models must be the source of hypotheses which dictate both the choice of logs pertinent to the problem and the types of mathematical methods to be used in their analysis. Most log analysts are busy in locating oil and natural gas reservoirs and in estimating engineering properties. The effective use of logs is common to all complicated applications, e.g., location and definition of stratigraphic traps, complex reservoirs, tertiary oil recovery, etc, and their geological information content is a crucial aspect. Quantitative analyses are preferable to qualitative interpretations, since these will form the basis of important decisions concerning exploration drilling programs and choice of methods for reservoir exploitation. Physics, mathematics, and geology are the fundamental facets of log interpretation.

Detailed Encyclopaedia

This encyclopaedia is arranged in alphabetical order. The detailed encyclopaedia is given belows: –

Density Log – The commercial density logging tool is a device to estimate the bulk density of formations within a borehole. The density tool typically consists of a radioactive source of gamma rays, e.g., cesium-137, which is mounted on a pad that is pressed against the formation by the tool. These constant energy gamma rays radiate into the formation and interact with the electron clouds of the atoms they encounter by crompton scattering, photoelectric effect, and pair production processes. As a consequence, there is a reduction in the gamma ray flux which is measured by a short-spaced and a long-spaced detector.

This reduction is proportional to the average electron density of the formation. The electron density measurements may be converted to apparent bulk density through considerations of simple atomic theory. The Z/A ratio of most elements is close to 0.5, where

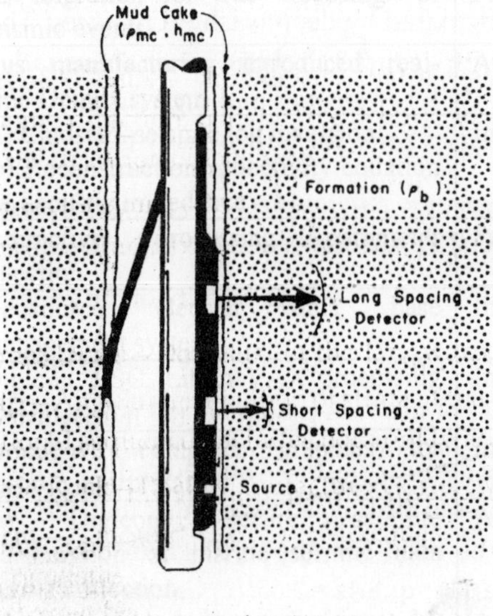

Fig. 1. Dual-spacing density tool.

Z is the atomic number and A is the atomic mass number. This is especially true for elements which are in the lower part of the periodic table, and it is these elements which occur in any appreciable abundance in nature. Since the number of electrons matches the number of protons, electron densities measured by the density tool may be translated to an apparent bulk density in grams per cubic centimetre or kgs per cubic metre, by using a Z/A ratio of 0.5. In logging typical sedimentary lithologies, the difference between apparent and real densities is trivial, since the real Z/A ratios for quartz, calcite, and dolomite are 0.4993, 0.4996, and 0.4994, respectively. But the density log apparent density for halite is 2.03 while its real density is 2.16, owing to its actual Z/A ratio of 0.4799.

The density log is conventionally presented as a trace of apparent density scaled between limits of 2 and 3 gm/cm^3. This range contains the densities of most lithologies encountered within a borehole, although a secondary scale of 1 to 2 g/cm^3 allows the recording of anomalous zones such as coal beds or excessive washout zones. Since the density tool is a "contact device", measurements can be adversely affected if the borehole is markedly rogose or punctuated by deep washout features. The log is run in conjunction with a caliper. Problem zones are usually obvious from their excessively low densities and high diameter readings on the caliper. For a clean zone, the bulk density is given by the following equation :

$$\rho_b = \phi\rho_f + (1-\phi)\rho_{ma} \qquad \text{.............. (1)}$$

Where ρ_b = Bulk density of the zone

ρ_f =Density of the pore fluid

ρ_{mu} = Density of the matrix mineral

Since the density tool has a limited radius of investigation, the pore fluid relates to the flushed zone rather than the virgin formation. Typical figures used for apparent grain densities of reservoir lithologies are (1) : quartz = 2.65, (2) Calcite = 2.71, and (3) dolomite = 2.87. When evaluating shaly reservoir units, the basic equation becomes as given below :

$$\rho_b = \phi\rho_f + (1-\phi-V)\rho_{ma} + v\rho_{sh} \qquad \text{.............. (2)}$$

Where ρ_{sh} is density of the shale and V is the proportion of the shale. Besides its applicatin to the estimation of porosity, the density log has a variety of uses ranging from the estimation of oil yields from oil shales to its application with sonic logs in the computation of impedance values for use in generating synthetic seismograms. It is a particularly valuable aid in the recognition of certain lithologies and minerals which have appreciable density contrasts with typical sandstones, carbonates, and shales. It is used widely in the evaluation of coal-bearing sequences both in the identification of lithologies and determinations of coal character, related to rank and ash content.

The nuclear reactions of gamma rays with borehole section materials is controlled to a large degree by energy level. At moderately high energies, interactions are dominated by crompton scattering, whose effect is determined by the number of electrons per unit volume and leads to conventional density measurements. However, at low energies, photoelectrical absorption becomes an important phenomenon and this is regulated by the number of electrons per atom, which is the atomic number, Z. Detector counts rates at low energy window can be used in the computation of a photoelectric absorption cross-section index, Pe, and recorded as an ancillary log in conjunction with the conventional bulk density trace. The index shows marked differences in value between minerals with the result that its log trace is a good lithology discriminator. Sandstones, limestones, and dolomites are generally readily distinguished, while the index is sensitive to the occurrence of minor amounts of heavy minerals.

Dipmeter – The dipmeter measures the orientation of bedding planes and other features within a borehole. Within a well any planar feature will intersect the borehole wall as an ellipse whose relative vertical displacement arround the hole is controlled by the dip and strike of the plane and the orientation of the borehole axis.

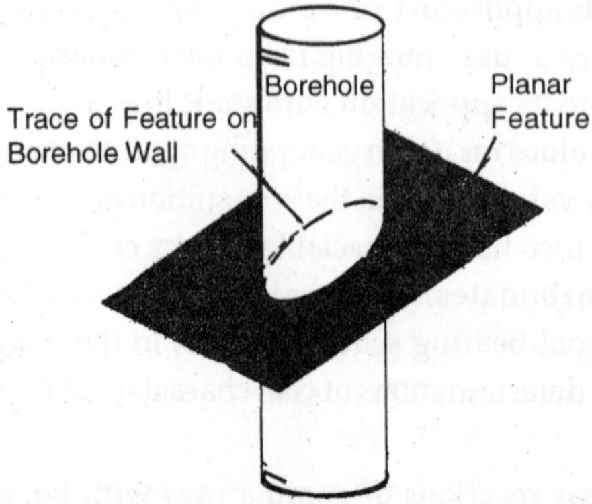

Fig. 1. Bore hole wall registration of dipping planar geological feature

A practical dipmeter must make measurements which have a fine vertical resolution. The modern dipmeter tool consists of either a three-arm or four-arm device, in which the arms are arranged radially around the central tool housing, and are linked by a spring mechanism which presses the arms against the borehole wall as the tool is run through the hole. On each arm there is a rubber pad in which are embedded small electrodes which make microresistivity measurements of the adjacent formation. The tool housing includes a compass and weighted pendulum which continuously monitor the geographic orientation of the arms and the deviation of the tool from the vertical.

The raw results of a dipmeter run are three or four microresistivity traces, together with the azimuth of one of the arms used as a reference and the borehole deviation angle. Since lithological changes are very minor when traced across the width of the borehole, the form of the resistivity traces is very similar. Major differences between them are expressed in a relative depth shift between correlative features, which is a function of their strike and dip. In most modern dipmeter runs, computer searches for a best match between resistivity segments on adjacent arms and computes the corresponding azimuth and dip. Three fundamental parameters control the operation of the computer program: (1) the window length, (2) search length, and (3) the step length. The window length determines the number of feet of the reference resistivity trace which will be used as a match with corresponding displaced lengths of traces from the other arms. The search length is the number of inches of maximum allowable displacement in the program's hunt for the best match on adjacent traces. The third parameter is the step length which determines the vertical distance in feet between successive correlation attempts.

In matching a segment of the reference trace with the resistivity trace from another arm, a statistical correlation coefficient is computed as a simple measure of match. When the coefficient is calculated at all possible match positions within the search length range, the maximum positive correlation is selected as the position of best match. If this best value fails to exceed some lowest acceptable limit, the computation at this position may be abandoned and new computations initiated at the next position determined by the step length. In the case of a set of satisfactory matches, the relative displacements between the trace defines the orientation of a plane relative to the borehole axis. The incorporation of the dipmeter readings of the borehole deviation and azimuth allows correction of the orientation of the feature with respect to a true horizontal plane. When penetrating high angle beds, the appropriate search angle will often dictate a greatly exaggerated search length. Similar problems will be caused in the analysis of highly deviated holes drilled from offshore platforms. The size of window length and step length should not be chosen independently. The window length obviously prescribes the lower limiting resolution of the maximum size of features that can be detected by the processing.The result is an averaging of dip information within the window. This is desirable in the extraction of structural dips. However, in the case of stratigraphic analysis, current bedding features will generally be lost both as a consequence of their thinness and dip variability. In dipmeter studies that are aimed specifically at the study of internal structures, the window length must be kept to a finer size. The increasing importance of stratigraphic studies in oil exploration has encouraged into the improvement of dipmeter analysis to fine detailed resolution. The presentation of the computed results is a "vector plot" or "tadpole plot". The vertical axis is depth and the horizontal axis measures the magnitude of dip on either an arithmetic or logarithmic scale. The location of each "tadpole" registers its depth of computation and calculated dip. The tail of the tadpole indicates the azimuth of the associated dip as related to a conventional compass circle. A colour convention is widely used by log analysts as a screening device to discriminate tadpole clusters which appear to represent significant common trends. Dip patterns can be variously attributed to tectonic dips, faults associations, compactional drapes and features internal to lithologies, such as cross-bedding.

Dipmeter Applications (Fabric Orientation Analysis)— The traditional application of the dipmeter is focused on the correlation of distinctive micro resistivity features across the borehole. By this means, local planes are established as surfaces with dip and strike. In topology, the data are restricted to increments of relative vertical shift, the orientation of the arms with respect to geographic north, and the deviation of the borehole from the vertical . If an individual feature shows the same resistivity value on all dipmeter arms, the rock fabric is suggested to be isotropic. If there is variation in resistivity between the arms, the fabric is an isotropic, and the degree of anisotropy is reflected in the range of its directional resistivity. Aggregate methods of fabric measure generally utilize variations in physical properties such as light intensity, imibition of water, acoustic velocity, dielectric constant and thermal, and electrical conductivity. These approaches are indirect, but have generally shown good correlations with observed grain orientations. The grain orientation of a sediment is obviously a function of the hydraulics of transport and deposition. The shape of most sand grains can be closely approximated by a triaxial ellipsoid. When transported in suspension, their long axes tend to parallel the current direction. In traction, the grains roll about their long axes with orientation at right angles to transport. The grains show imbrication upcurrent and this is the necessary key to specify current direction rather than merely axis of movement. On a three- arm dipmeter a single feature will be recorded as three different resistivity readings whose orientations are separated by 120 degree in the depositional plane. These three resistivity vectors define the trace of a resistivity ellipse as a realization of a model resistivity variation around the entire azimuth cycle. Dipmeter studies of shale fabric patterns offer useful indications of neighbouring faults and folds, whose influence could be mapped from dipmeters run in several wells.

Dipmeter Interpretation (Sedimentary Structures and Depositional Environments)—The scale of sedimentary structures contained within stratigraphic units is usually drastically smaller than forms associated with tectonic folding, faulting, and elements of significant compactional draping. Most cross beds are less than 1 foot thick. There is the common tendency for cross-bedded surfaces to show complex polymodal patterns of variable dip and orientation. Eolian cross beds cause extremely distinctive and coherent patterns to be registered on the dip vector plot. They are easily discriminated in computer processing. Blue patterns characterize the toeset, with low angles close to structural dip at the base. These patterns merge upward into high angle (around 30 degree) green patterns in the dune fore-set which match the angle of repose. The high consistency of dip orientation within each dune set is characteristics of colian deposits, but there is often a marked variation between sets.

Braided stream deposits are produced by complex networks of channels which carry high loads of gravels and coarse sands. They are usually found in areas undergoing significant rates of erosion and form thick and extensive sheetlike bodies which often overlie an irregular surface of unconformity. A dipmeter profile of a braided stream sequences generally shows a highly variable set of patterns which are impersistent and erratic. The small diameter of the borehole results in a dipmeter perception of nonplaner surfaces and localized lenses of internally coherent dips. White patterns are commonly observed, while limited red and blue patterns pick out through cross beds and erosional surfaces. Shales formed in abandoned channels may register green patterns which indicate the general direction of structural dip. Within the coarse clastics, dips can vary from horizontal to angles as high as 35 degree. The foresets show variability in bearing, but are approximately contained within a 90 decree arc whose mean orientation gives an indication of the downstream direction. This trend should roughly parallel the axis of elongation of the sandstone body.

Fluvial sediments are finer grained than braided channel deposits, with high proportions of both sand and shale. Both outcrop and cored intervals show a dominant motif of upward fining grain size sequences. These range upward from a channel floor erosion surface, which may have an intraformational conglomerate, through a succession of cross-bedded point- bar sands into cross-laminated fine sands and overbank shales. The meander of the channel causes the stream bed to migrate laterally, with accretion at point-bar surfaces. Within the point bar, trough cross-beds of relatively coarse sands grade upward into thinner tabular sets with finer sands. Abandoned channel loops are filled with deposits of silts and clays. The meander belt deposit forms a sinuous body which may be up to 20 times the width of the stream. Dips are generally observed to be highly inclined and somewhat erratic at the base of trough cross-bedded units, but become more coherent and shallower in the overlying tabular cross-beds. As a consequence, downward increasing red dip patterns typify channel sections. Detailed analysis often reveals a bimodal pattern of orientation. Red patterns in the overlying shales suggest compactional draping features. Green patterns within the shales discriminate the local structural dip.

The meandering streams of the upper alluvial plain pass into distributary channels of the delta. These channels are less sinuous in plan and form natural levels of clays and silts. The width of the channel fill is generally much less than those of alluvial meander belts, so that the deposits form long narrow bodies. Delta distributary channels can show complex variations in dip pattern. However, the basic motifs follow similar patterns to those found in meandering stream point bars. The morphology of deltas is controlled primarily by the sediment load and the balance between fluvial and marine regimes. A spectrum of delta shapes are possible according to the strength of destructive wave processes relative to constructive fluvial deposition, e.g., birds foot, through lobate, arcuate, and cuspate forms. Highly constructive deltas form lobate sand bodies at the mouths of distributaries and extend seaward. While relatively fine grained at the base, there is a general trend of a coarsening upward profile. At the same time, the cross-bedding angles are generally steepest at the top and shallow downward to cause distinctive blue patterns on the dipmeter vector plot. The cross-beds are normally tabular in form and inclined seaward, unless reworked or influenced by long shore currents. As the delta progrades seaward, the feeder channel may cut down through the upper part of the bar.

In cases where deltas are dominated by strong tides there is oscillating interplay of fluvial and marine processes. At flood tide, seawater moves upstream into estuaries and tidal channels, but recedes in time of ebb tide. The net effect often causes a bimodal pattern of diametrically opposed dip directions which are picked up by the dipmater. Otherwise, the internal characteristics of these sands resembles those of distributary channel fills when deposited in moderately narrow estuaries. Wider estuaries generally develop extensive tidal flats as silt and mud complexes cut by subparallel sand bars. The grain size profile typically shows a ragged fining-upward gradation from coarse sands at the base to shales at the top. On coast lines with strong marine currents and a source of sediment, linear shoreline elements are formed parallel, to the coast. These include both beaches and offshore bars and islands found at or close to wave base. Most beaches show upward coarsening in grain size and tabular cross beds which dip towards the sea. Submerged offshore bars show characteristics similar to beaches on their seaward side. Dips on the landward side slope in the opposite direction and are often greater than those on the seaward flank. With sufficient sediment supply, the bars may emerge and form a barrier complex with lagoons or swamps behind them. Dips are characterized by downward-decreasing blue patterns. At flank positions, the dips are generally unimodal in orientation and reflect either shallow seaward-dipping beds or the steeper forsets on the landward side. Close to the bar axis, the patterns may be bimodal and diametrically opposed dips. Dips are arranged at right angles to the bar axis. The upper surface of the bar is generally convex upward and causes compactional draping in the overlying shales. Many marine sands are sheet-like in form with repetitive regressive transgressive units. Dips may form localised blue or red patterns but are often polymodal or random in orientation as broad white patterns.

Dipmeter Patterns (Structural Geologic Interpretations) – The recognition and analysis of geometrical patterns are considered from dipmeter data that reflect structural features on all scales, e.g., regional dip, local structure, faults, unconformities, and the results of draping caused by differential compaction of stratigraphic units within a sedimentary sequence. Since they are generally large in scale, they are most easily extracted from computer processing of dipmeter logs. As a general rule, structural dip will show as monotonous but easily discremible green patterns which collectively form a long-term trend. Other patterns usually occur as localized motifs which match the limited vertical scale of the geological features that control them. The recognition of unconformities on the dipmeter log is mainly contingent on the geometrical relationship between beds above and below the erosional or nondepositional surface. Discrimination is obviously easiest in an instance of a pronounced angular unconformity. However, similar breaks in structural dip can be caused by faulting. Some knowledge of local geology is crucial to intelligent interpretation.

In order to be detected on a dipmeter log, faulting must cause bedding plane distortions about the fracture zone or result in differential tilting of the fault blocks. Growth faults are normal faults which are formed contemporaneously with deposition and often show greater formation thicknesses on the downthrow side. They commonly develop "rollover" features which provide local anticlinal traps for hydrocarbons. The resulting distortion of bedding planes in the hanging is easily detected by the dipmeter and is generally contained within a 10 foot interval above the fault plane. Drag faults are generally formed after deposition, but at a stage before complete compaction of the sediments. The poor degree of mechanical competence results in a relative local warping of the hanging wall into the fault plane with effects registered by the dipmeter. In more consolidated, and therefore competent, sequences the effect of a normal fault may go undetected in the dipmeter log if the result is one purely of fault block displacement with no distortion. Similar observations are true for thrust or reverve faults, although the stress mechanics of thrust faulting commonly result in bed rotations in both hanging and foot walls.

Stratigraphic sequences typically consist of beds with varying degrees of mechanical competence. When subjected to burial and the associated pressures of lithostatic loading, a sediment pile exhibits differential degrees of compaction, both vertically and laterally. If relatively rigid lensoid bodies are surrounded by less competent units, originally planar surfaces will be distorted in local structural features of compactional draping. The dipmeter can be used to pinpoint the location, shape, and orientation of nearly barrier sands, channels, and carbonate reefs. The dipmeter is useful in the detection of potential reservoir units which are not actually encountered within the well bore. Barrier bar sands generally parallel ancient shorelines and have a convex upper surface. Differential compaction of bounding transgressive shales results in a draping of beds over the crest of these features. In wells drilled at flank positions, the drape is shown in the shales overlying the sandstones as red patterns, in which dips diminish upward to overlying green patterns representing local structural dip. The crest of the bar is therefore located updip from the pattern trend and the bar axis parallels their strike. Channel sandstones form elongate bodies. Their cross-sectional shape varies according to the geological context, although almost all have a convex-downward erosional base. When the channel has cut down into compacted lithologies, the upper surface may be broadly concave-upward in form. The draping of overlying shales is effectively a synclinal features which dips inward to the sandstone axis. The resulting dip sequences in the shales of flank positions are red patterns which specify the direction to the axis and its local orientation. When the channel is cut into soft beds such

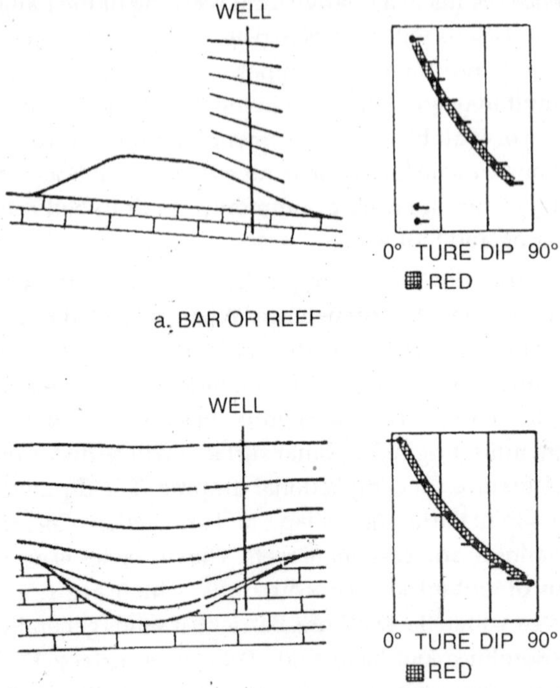

Fig. 3. Dip patterns associated with draping over a barrier (a) Sand bar and (b) A channel sandstone (After Campbell, 1968)

as shales, the compaction of the sequence generally results in a biconvex cross-section as the profile of the sandstone body. This situation causes draping characteristics in the shales overlying the channel sand similar to those of a barrier bar sand. Dip interpretation of draping is specific to the shape and orientation of sandstone bodies, rather than diagnostic of depositional environment. Reefs of limestone and dolomite form surface features with a relative relief that may reach hundreds of feet. When their adjacent and overlying facies are marine shales, the compactional draping effect over the reef flanks is highly pronounced.

Dipmeter Vectors (Analysis)—All sources of information must be integrated with the dipmeter log, including the analysis of other logs, cuttings, and core discriptions, and convex within regional geology. Once the appropriate model has been identified, dipmeter vectors can be analyzed to extract detailed measures of the three-dimensional configuration of both structural and sedimentological elements. The results of the dipmeter survey are presented in a variety of forms : (1) a listing of the dips and azimuths at various depths, (2) a vector or tadpole plot of the same information, and (3) a cylindrical plot. All three forms present alternative views of the same dipmeter information. More detailed graphic analysis may be made of subsections of the whole interval through the use of polar plots. The most commonly used variants are the Schmidt net or the Wulff net. On either project, the normals to the dipping planes are plotted with respect to a hypothetical lower hemisphere or the upper hemisphere. The Schmidt net presentation has the advantage that equal areas on the hemisphere surface will be preserved as equal areas on the projection. The Wulff net preserves great circles on the hemisphere as great circles on the projection.

The dip vectors can also be analysed through the use of azimuth frequency diagram (rose diagrams). In common with dip vector plots, and polar plots, this method is purely graphical in the interpretation of trends and relative dispersions. The correct approach to the analysis of directions in three-dimensional space lay in statistical analysis of vectors summed in the form of an "orientation tensor". The principal eigenvector of this orientation tensor specifies the mean direction of the component vector. The role of eigenvalues is on the characterization of the overall fabric shape which characterizes the set of vectors taken as a group pattern of orientation. The preferred orientation of the dip plane normals in the Schmidt net plots may be numerically analyzed in terms of their distribution on the reference sphere. The eigenvectors of orientation tensor specify the major, intermediate, and minor axes of the distribution of normals, and are identified by the size of their respective eigenvalues. Each eigenvector contains the direction cosines that specify its orientation. Each eigenvector corresponds to the proportion of the total variance which is accounted for by its associated eigenvector. A "scattering angle" θ of the normals about the principal axis may be computed by the formula given as :

$$\theta = \text{arc}\cos\sqrt{\lambda} \qquad \text{................ (3)}$$

Where : λ is the eigenvalue of the principal eigenvector.

Gamma Ray Log—Gamma rays have significantly larger penetrations and can be detected by simple counter devices. Only three types of radioactive isotopes occur in any appreciable abundance in natures, namely: the uranium series, the thorium series, and the potassium-40 isotope. Scintillation detectors are used in gamma ray tools. This tool can measure formation characteristics through casing. The selection of an appropriate logging speed and time constant is made as a compromise which seeks to minimize the component of statistical noise, while not eliminating the effect systematic minor variations in radiation as part of the averaging process. On each well, a check is made of the statistical background fluctuation by holding the tool stationary within the borehole for a few minutes. A trace of this statistical check is generally recorded on the log to be used in the evaluation of minor anomalies. The gamma-ray tool is field calibrated at each well by exposing the tool to a standard calibrator at a fixed distance. All modern logs are scaled in terms of the API gamma-ray unit. In the majority of stratigraphic and petroleum geological applications, the gamma-ray log is used as a "shale log," both to differentiate shales and "clean" formations and to evaluate shale proportions in shaly formations. The general form of the gamma- ray log often shows a broad similarity with an SP log run for the same section, with shales to the right and clean formations to the left of the log-track.

A "shale baseline" is drawn which is a best estimate of the reading for "normal shales" in the sequence. Similarly, a "clean formation line" is drawn, which is roughly coincident with the lowest readings that occur in clean units of interest. In each case, these lines would be drawn to satisfy a general plateau of values which appear to correspond to natural boundary levels. If the clean formation and shale lines have API values of C and S respectively, the "gamma-ray index" (GRI) of any zone, with associated reading, G may be estimated by the following equation :

$$\text{GRI} = \frac{G - C}{S - C} \qquad \text{........... (4)}$$

This equation represents a simple linear interpolation between the two boundary values. The GRI is an appropriate measure of shale volume on an arithmetic scale. In practice, the clay mineralogy, volume, and nature of silt content will generally vary between thick shales and thin shales within sandstone and carbonate beds. This variation reflects the changes of clay mineral associations that characterize different depositional facies and the results of diagentic changes. The GRI is an empirical measure of shale volume content. It generally provides an acceptable estimator which is useful in stratigraphic studies and for mapping purposes. Most of the current log analysis techniques can only provide estimates of shale content rather than specific clay mineral content. Gamma-ray log profiles may be applied for the interpretation of depositional environments of both sandstone shale and carbonate-shale sequences. The gamma-ray trace records the nature of upper and lower contacts as sharp or gradational, and gives indications of shale content variation. The gamma-ray can be used in the distinction of uranium rich black shales and possibly in the recognition of arkosic sandstones and zones of high heavy mineral concentration.

Gamma-Ray Log (General Applications) – Chemical analyses and work with the gamma ray spectral log have been useful aids in tagging the sources of radiation within various lithologies to specific isotopes. The clay minerals of shales contain small amounts of potassium, of which 0.02 per cent is the radioactive isotope K- 40 and may account upto 20 per cent the total radiation emitted. More important are concentrations of thorium which appear to be fixed to the clay minerals. Small amounts of uranium also occur in almost all shales. In certain shales uranium is concentrated in high proportions which is generally a reflection of depositional conditions of highly reducing environment. High uranium shales are particularly useful as stratigraphic markers for correlation purposes. Most carbonates show very low levels of radioactivity unless they contain disseminated shale or have been mineralized by uranium bearing solutions. Simple orthoquartzites show low values unless significant amounts of shale, felspar, mica, or zircon are present.

Appreciable variations of radioactivity in sandstones have traditionally been attributed to shale content whose magnitude is computed in order to make necessary corrections to porosity and resistivity readings. The cation exchange of contaminating clay minerals tends to reduce the resistivity reading and must be corrected for in water saturation calculations. Micas are basically insulators. Gamma-ray spectral logging runs have indicated that the major source of radioactivity in typical sedimentary micas is a moderate amount of potassium, with only very minor contributions of uranium and thorium. The gamma-ray tool is widely applied in uranium exploration. This tool is useful in the search of minerals with high potassium content, e.g., sylvite, other salts in evaporite sequences. When appropriate calibrations of the tool are made, direct translation of the API count scale to equivalent concentration of potassium oxide is possible. Gamma-ray logs are commonly used for stratigraphic correlation in the subsurface. A comparison of an outcrop radioactivity profile with subsurface gamma-ray logs are possible. Gamma-ray scans of outcrop should be standard part of every description of a shale.

The spectral gamma-ray log can be used in the geological application of subsurface radioactivity measurements. The spectral device makes use of the fact that gamma-rays from potassium-40, uranium, and thorium sources are characterized by distinctly different energy level. Fig.4. By subdividing the total energy range into restricted intervals or windows, estimates can be made of the separate contribution of potassium-40, uranium, and thorium sources. The process of radioactive decay is stochastic. Consequently, statistical estimation theory must be applied to produce coherent values which reflect real variation rather than statistical counting error. The results are recorded as a spectra1 gamma-ray log which shows three separate traces referred to a radioactive -f source as a junction of depth . The relationships between radioactive elements and geological hosts have been studied, based on both subsurface and laboratory measurements. Interpretations are sometimes complex or even ambiguous, since the emplacement of radioactive

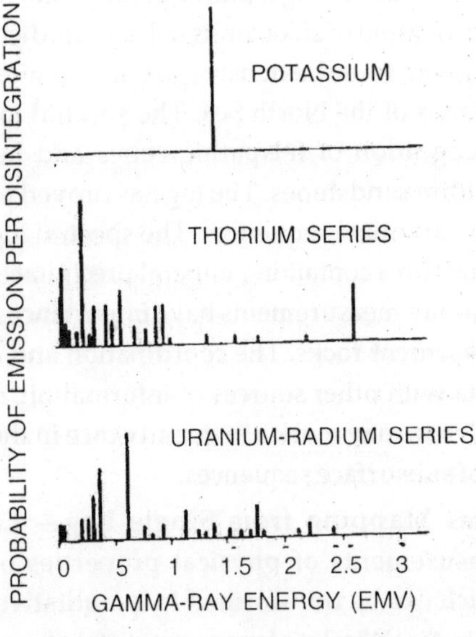

Fig. 4. Gamma-ray emission spectra of radiative sources (After Serra et. al. 1980)

materials is the result of a variety of mechanisms, acting singly or in concert, e.g., depositional environment controls, diagenetic effects, etc.. The original source of potassium is chiefly in acid igneous rocks within potassium felspars, micas, and some minor accessory minerals. Thorium and uranium are also derived from acid igneous rocks. Most organic shales tend to be anomalously radioactive.

Gamma-ray spectrometry of outcrop samples is a useful method to evaluate shales in terms of organic carbon content, paleo-environmental interpretation, and stratigraphic correlation. Variations in the ratios of gamma- ray sources could be correlated for hundreds of kilometres across lithofacies boundaries. Results could be used to deduce sediment transport directions, suggest proximity of paleoshorelines, and aid in the interpretation of lateral and temporal changes in paleosalinity. Direct comparison could be made between outcrop profiles with gramma-ray spectral logs of equivalent units in the subsurface. The combinations of outcrop spectral data and spectral logs shows great promise for detailed geological interpretation of shale units in basin studies. The spectral gamma- ray log has found extensive use in the interpretation of Jurassic sandstones of the North Sea. The spectral log also allows the specific recognition of felspathic zones and heavy mineral concentrates within sandstones. The log has proved to be a distinct asset in logging carbonate sequences. The spectral log can be used to distinguish fractures containing mineral precipitates from shales. Spectral gamma- ray measurements have importance in the logging of crystalline basement rocks. The coordination and calibration of spectral log data with other sources of information, such as X-ray diffraction analyses, may mark a major advance in the detailed log interpretation of subsurface sequences.

Lithofacies Mapping from Single Log – Log readings are direct measurements of physical properties of subsurface formations which can be transformed to quantitative estimates of geological attributes. Wireline logs are remotely sensed data from the sub-surface. Interpolation between a network of well control will generate a spatial construct whose form is a reflection of geological variability. Log analysis of lithofacies must first be consistent with the geology of the formation found by drill bit. At another level, the less tangible rock characteristics of palentology, sedimentology, and diagenesis are the necessary keys to provide genetic meanings to patterns derived from logs. Most of the commonly available logs are primarily sensitive to variations in the volume of pore fluid and/or the fractional content of shale when run through typical sedimentary sequences of limestones, dolomites, sandstones, and shales. The mapping of porosity trends is obviously not a direct expression of lithology. The simplest application of lithofacies mapping from a single log is the evaluation of lateral shale content across an area. The gamma-ray log is the most insensitive to porosity variation. Gamma-ray log values may be transformed to profiles of estimated shale content by interpolation between limits that appear to typify shales, and sandstones or carbonates as the scale boundaries. The rescaled gamma ray trace expresses shale content and its vertical variation as a fluctuating distribution which must be condensed in some manner to descritors useful for mapping. The most basic measure is the mean shale proportion, which reflects the aggregate composition. Interpolation of mean shale values of a formation between wells results in the creation of continuous surface of shale variation across geographic space. The form of the surface reflects lateral changes in the gross aspects of shale geometry.

The selection of gamma-ray readings that set shales and sandstones or carbonates as scale boundaries, result in a transformation of the log into a shale profile. This new scale is keyed to estimated proportion shale content rather than arbitrary radioactivity count units. Through use of scale, the average shale composition can be computed. If the log is digitized at an appropriate depth interval, then the mean shale proportion $\overline{S}$ is given by the equation: –

$$\bar{S} = \sum_{i=1}^{n} \frac{Si}{n} \qquad \text{................ (5)}$$

where Si is the shale proportion at depth i and the sequence contains n digitized values. The center of gravity V_1 is computed, from the following equation :-

$$V_1 = \frac{\sum Sidi}{\sum Si} \qquad \text{............ (6)}$$

Where di is the depth of the ith value. The centre of gravity is the first movement of the shale variation. The second moment of the shale variation is given by the equation: –

$$V_2 = \frac{\sum Sidi^2}{\sum Si} \qquad \text{.......... (7)}$$

A progressive series of moments may be calculated from the general equation given below: –

$$V_m = \frac{\sum Sidi^m}{\sum Si} \qquad \text{.........(8)}$$

where V_m is the mth moment. Maps of these separate moment characterization are made in order to assess the variation of the contained shale in both the vertical dimension as well as the two lateral dimensions of geographic space. Interpretations from the isopach and three shale statistic maps can be amplified by more detailed comparison of the separate maps or examination as overlaid sets. A set of movement maps defines three-dimensional variation in a precise quantitative manner. Moments are the fundamental parameters of polynomial regression curves.

Conventional moment maps represent surfaces of variation generated by interpolation of moments computed from logs at well control. When interpolated estimates of moments are inserted into the matrix algebra solution for a polynomial regression, the solved coefficients define a curve of vertical variability as a projection of trends from wells in the immediate vicinity. Since polynomials are continuous functions and it is possible to specificy an infinite variety of geographic locations, there is no theoreti cal limitation to the prediction of a shale trend value at any combination of geographic coordinates or depths. Polynomial functions provide useful parameters for mapping. A polynomial equation is given below: –

$$\bar{S} = a_0 + a_1 d + a_2 d^2 + \text{..........} + a_m d^m \qquad \text{........(9)}$$

where $\bar{S}$ is the trend value and d is the depth. Differentiating equation (9), we have :

$$\frac{d\bar{S}}{dd} = a_1 + 2a_2 d + \text{...........} + m a_m d^{m-1} \qquad \text{.........(10)}$$

which gives the slope of the polynomial at any depth. The location of the peaks and troughs of the polynomial can be extracted when the slope equation (6) is set to zero and solved for depth. Differentiating equation (6) gives :

$$\frac{d^2\overline{S}}{dd^2} = 2a_2 + \ldots\ldots\ldots\ldots + m(m-1)a_m d^{m-2} \quad \ldots\ldots\ldots(11)$$

which is an expression of the rate of change in slope. The ratte of change is zero at curve inflection points, which mark the boundaries between peak and trough features. The solution of inflection points at successive grid column can be linked to sketch out subdivision boundaries. The result is a method of polynomial stratigraphy and is a mathematical solution of lithostratigraphy based on the trend differentiation of major subdivisions. The transformation of moment maps to three-dimensional representations is not confined to the generation of vertical slices matched with cross-sections. Horizontal slices can be produced which portray trend estimates of shale proportion at any specified depth.

Three-dimensional trend mapping not to be restricted to processing of gamma-ray logs. Three-dimensional characterizations of trends in porosity within reservoir units may be produced using the same methodology applied to neutron, density, or sonic logs. The major assumption would be that the range of lateral continuity in porosity trends exceeds the spacing of control wells. An interesting consequence of three- dimensional mapping of sonic logs on cross-section slices is that they should show strong similarities with synthetic velocity profiles generated from the deconvolution of seismic records. The method may find useful applications in seismic modelling through studies which integrate processed seismic records and regional three-dimesional velocity trends based on sonic logs. The possible polynomial equations can be considered as a set filters in which low-order polynomials are low-pass filters. At progressively higher orders, the polynomials pass increasingly higher frequencies. In general, low frequency information shows greater lateral continuity than do high frequencies, which often represent variation caused by thin, impersistent beds. However, the ultimate controls are the thickness of the analytical section, its vertical variability, and the spacing distance between control wells.

Lithofacies Mapping (Complex) from Multiple Logs – More detailed assessments of compositional variation may be made if several logs are studies in combination. The operation requires a computer contouring program that can handle grid-to-grid manipulations in addition to surface estimation and map production. The mapping of changes in lithology across an area requires an extension of the methodology developed for individual well profiles to techniques suitable for lateral interpolation. In order to compute the unit mean composition at any location, the basic input requirement is merely the average log responses of the unit rather than a complete suite of log traces. Since each average log response of a major stratigraphic unit will fluctuate only gradually when traced laterally between wells, these variations will sketch out continuous hypothetical surfaces across the area. Such surfaces are generally approximated by computer mapping packages as grids of numerical values. Grid rows and columns specify geographic position, while the grid mode values are estimated by interpolation from data in available well control. Core and cuttings are the source of information which puts flesh on the bones of variability mapped from a log. Bodies that are mapped with low shale contents can be assigned specific identities as either sandstones, limestones, dolomites, or other low-radioactive lithologies. These distinctions are not recognized explicitly by a single log. Mapped variations in porosity can be interpreted in terms of distinctive lithofacies. Once again, it is the analysis of core and cuttings that is the supplementary key which provides geological meaning.

Lithology Estimation (Numerical Methods)—Numerical techniques can handle any number of logs and provide immediate quantitative solutions which are cumbersome to read from graphs. The mathematics for lithofacies analysis from logs may be drawn from simple matrix algebra. From an operational standpoint, matrices are easily stored and manipulated in a computer and most programming languages are structured in terms of a matrix representation. Consequently, lithofacies analysis programs can be written which are cheap and fast to run, even for extensive logged sections. A matrix algebra solution of component compositions is equivalent to an eyeball compositional estimation of crossplotted zones. At the most basic level, the geometrical implications of matrix algebra data provide some means of visualizing the relationship between lithofacies compositional solutions and the original log data. This method can be used with any logs that are sensitive to compositional content in any system of components regardless of their origin.

The basics of compositional analysis is matrix algebra. If knowns are C and L, and unknown is V, they are related by the equation given below :

$$CV = L \qquad \text{...............(12)}$$

$$V = \frac{L}{C} \qquad \text{..............(13)}$$

In matrix algebra where each letter signifies an array of numbers or unknowns, rather than a single number or unknowns, as in conventional algebra. The equivalent operation of division is not feasible in matrix algebra. In matrix algebra equation (13) will be written as :

$$V = C^{-1} L \qquad \text{.............(14)}$$

where C^{-1} is a inverse matrix. The inverse matrix has the basic property given in the equation as :

$$C^{-1} C = I \qquad \text{............(15)}$$

A matrix multiplied by its inverse matrix gives an identity matrix I, in which all the diagonal entries are "ones" while the off-diagonal entries are zero.

Matrix algebra can be used for compositional analysis of logs. Wyllie et al (1956) concluded that in pure lithologies with small pores, transit time measured by sonic logs is closely approximated by a linear function of porosity. This relationship is most commonly expressed by the "Wyllie time-average equation" which is given below: –

$$\phi = \frac{\Delta t - \Delta t_{ma}}{\Delta t_f - \Delta t_{ma}} \qquad \text{............(16)}$$

Where ϕ = Porosity of the zone

Δt = Transit time of the zone

Δt_{ma} = Transit time of matrix mineral (zero porosity)

Δt_{f} = Transit time of the porosity fluid (100 per cent porosity)

Equation (16) is within the expanded form, this becomes the following equation set :

Sonic : $\phi.\Delta t_f + p_{ma}\Delta t_{ma} = \Delta t$ ....(17)

unity, $\phi + p_{ma} = 1$ ....(18)

In matrix form, this is :

$$\begin{bmatrix} \Delta t_f & \Delta t_{ma} \\ 1 & 1 \end{bmatrix} \begin{bmatrix} \phi \\ p_{ma} \end{bmatrix} = \begin{bmatrix} \Delta t \\ 1 \end{bmatrix} \quad \text{............. (19)}$$

or more simply as

$$CV = L \quad \text{............. (20)}$$

In a lithology of clean but variable composition of dolomite and calcite with small pores, and additional log, such as a density, would be required for explicit solutions. The descriptive set of equations would be:

Sonic: $\phi.\Delta t_f + p_d\Delta t_d + p_c\Delta t_c = \Delta t$ ....(21)

Density: $\phi\rho_f + p_d\rho_d + p_c\rho_c = \rho_c$ ....(22)

Unity: $\phi + \rho_d + \rho_c = 1$ ....(23)

Which in matrix notation can again be symbolized as:-

$$CV = L \quad \text{..........(24)}$$

although the size of the matrices are expanded. Solution of the composition of zones logged through such a sequence could either be solved graphically by recourse to a sonic-density cross-plot, or numerically by a matrix algebra solution.

If the problem was further expanded to clean, cherty, dolomitic limestones for resolution by measurements from a neutron-density-sonic log unit, there would be four equations in a set. A graphic approach to the solution would require a projection of a three-dimensional situation onto two-dimensional paper which would be considerably time consuming. Using the matrix algebra approach, the computational requirements are a single inversion of the coefficient matrix (which preceds the reading of any logs), followed by one matrix multiplication operation for each zone. In all cases, n components will require (n –1) logs for an explicit solution. Typically, these logs will be drawn from either /and/or the neutron density, and sonic logs, coupled with the gamma ray, and occasionally, the spontaneous potential or a transformed resistivity log.

The matrix formulation is a linear model and generally provides a satisfactory approximation for compositional solutions, as confirmed by core analyses and laboratory studies. Its basic limitations is that the number of logs should match the number of components minus one. In this situation, the equations match the unknow and the system is said to be "determined". In reality, this condition will not always be met. In many instances, the number of logs will be insufficient to provide unique solutions of the compositions of complex lithologies. This is cabled underdetermined systems. Less frequently, the number of logs may be more than necessary to resolve lithologies with simple mineralogies.This is called over determined systems. The underdetermined system is solved by the "least error" method. The proportional variance is given by the following equation :

$$P = \frac{n}{n-1} \sum vi(1 - vi) \qquad \text{.............(25)}$$

where V_i the proportion of component *i*, subject to the constraints of the information supplied by the available logs. The proportional variance can be re-written as: –

The solution corresponds to

$$P = \frac{n}{n-1}\left(1 - \sum V_i^2\right) \qquad \text{............(26)}$$

The solution coresponds to

$$v = C^T (CC^T)^{-1} L \quad \text{.............(27)}$$

where L is the vector of log responses C^T signifies the transpose of the G matrix, which simply means a matrix in which the rows and columns have been interchanged . In no log information, use of the unity equation gives the component vector estimate as

$$V = \left[\frac{1}{n}\right] \quad \text{.............(28)}$$

If several logs (not sufficient) are available, the equation will generate compositions which are the nearest feasible solutions to a homogeneous mixture of the components. It assumes that the entire range of possible compositions are equally likely but this is generally not the case for real rocks. High proportional variance solutions need additional geological information and this can be incorporated within the matrix algebra. The minimize variance is related to a new reference point which is more likely to represent the real composition of the section. A "guide point" composition can generally be estimated, based on regional stratigraphic considerations or studies of nearly outcrop or cores from neighbouring wells. The modified "guided minimum variance" solution is then

$$V = C^{T.} (CC^T)^{-1.} (L+CD) - D \quad \text{.............(29)}$$

$$D = N - Z \quad \text{.............(30)}$$

Where N is the vector of proportions, l/n, and Z is a vector of the guide point proportions. The method has useful feedback information. When the number of logs is less than the number of components by one, the equations satisfy the unknown and the appropriate matrix algorithm is a conventional simultaneous equation solution procedure.

Then:-

$$V = C^{-1} L \quad \text{.............(31)}$$

This is the most common matrix method that is used for the calculation of porosity and mineral-component proportions.

In overdetermined system an appropriate estimate of zone composition can be drawn from a least-squares model in which the error is minimized between the log response and its corresponding value predicted by the solution. The matrix solution of the least-squares model then becomes:

$$V = (C^T C)^{-1} C^T L \qquad ....(32)$$

The vector of errors, $\bar{E}$, associated with the solution for each zone may be computed readily as

$$\bar{E} = (-L : C)\begin{bmatrix} 1 \\ \overline{V} \end{bmatrix} \qquad ....(33)$$

and a mean absolute deviation, e, is given by

$$e = \frac{\sum_{j}^{m} | e_j |}{m} \qquad ....(34)$$

where *m* is the number of logs.

A coordinated solution procedure is given below :

The matrix algorithms described form a coherent hierarchy of equivalent operations in the solution of porosity and mineral components in lithologies from the analysis of log suits. Their main features are summarized in Table l.

In each situation, the appropriate procedure follows a similar computational sequence. A system of components is selected on the basic of prior geological knowledge or from inspection of log crossplots of the zones. A matrix of component log responses is specified a prior for the designated components from available laboratory measurements of their physical properties. The match of the number of components with the number of logs specifies the degree of determinancy of the solution system. The appropriate matrix operations of transposition, multiplication and inversion then are applied to the coefficient matrix as a single execution which precedes the processing of the logs. Component proportion solutions for any zone may be obtained immediately by the postmultiplication of the vector of log responses and unity. When implemented in an

Table.1

Matrix Alogithms for Differennt Contingencies Relating Number of Component Unknowns with Number of Available Logs.

Number of Component	Number of Logs	System	Solution	Alogrithm	Error Diagnosis
n	<n–1	Underdetermined	Maximum variance	$V=C^T(CC^T)^{-1}L$	*p, a*
n	<n–1	Underdetermined	Guided min.	$V=C^T(CC^T)^{-1}X-D$	*p, a*
n	n–1	Underdetermined	Unique	$V=C^{-1}L$	*p, a*
n	>n–1	Overdetermined	Least squares	$V=C^T(CC^T)^{-1}C^TL$	*p, a, e*

Fig.16. Basic descriptive shapes of SP profiles.

interactive environment of a computer, the simplicity of the algebra allows rapid processing of large volumes of log data. As a consequence of the solutions and their associated diagnostics, the user may respond immediately with modifications at all levels, e.g., components, coefficients, stratigraphic intervals, logs, in an interactive strategy to isolate solutions which are most consistent with the available data.

Log Trends and Patterns (Mathematical Analysis)– Useful interpretations may be made from visual examination of raw logs, log crossplots, and processed compositional profiles, when these are coordinated with available core and cuttings. However, the numerical form of log data makes Iogs an ideal medium for quantitative analysis. Log data are both numerica1 and numerous. The field of mathematical geology has drawn on theory from statistics, signal theory, pattern recognition, and other areas in adaptation of methods for the numerical analysis of geological variations. Mathematical techniques are applied to logging data as aids in the interpretation of geological patterns. The isolation of systematic relationship between individual components may give insight into facies associations and diagenetic mechanisms.

The depth axis of any log is some monotonic function of geological time in sequences which are not faulted or overturned. As a result, most log traces are a natural expression of a time series of geological variation. Any systematic trends that are extracted should reflect major stratigraphic patterns in the disposition of limestone and charty dolomite facies. The application of the methodology can be illustrated in the analysis of calcite variation as a function of depth.

A first order polynomial is given by the equation :

$$c = a_0 + a_1 d \qquad ....(35)$$

where c is the trend value of caicite content and d is the depth, in an equation which defines a straight line. The second order polynomial is

$$c = a_0 + a_1.d + a_2.d^2 \qquad ....(36)$$

It described a parabola, a curve with a single maximum or minimum. The general equation of an *n*th order polynomial is

$$c = a_0 + a_1.d + ... + a_n.d^n \qquad ....(37)$$

This will have n–1 maxima or minima and n–2 inflection points, expressing peaks/ toughs and boundaries of subdivisions, respectively. Conventional methods of zonation used in basic log analysis are easily duplicated in the subdivision of polynomial trends. Individual log zones are usually identified with distinctive peak or trough features on the log trace. Boundaries between adjacent zones are located at the position of inflection points on the curve shoulder which links the two zone extremes. A polynomial equation may be solved for depths matched with maxima, minima, and inflection points by simple method of calculus. The use of calculus on a polynomial trend function is therefore a simple method for stratigraphic subdivision, where inflection points locate boundaries, while extremes mark the zenith or nadir of the fitted variable within each zone. The criteria of this "polynomial stratigraphy" differs philosophically from conventional stratigraphic methods. Instead of mathing stratigraphic horizons with discrete marker beds or distinctive breaks, the trend curve geometry accommodates the total variation of the unit.

The simplest method to smooth fluctuating data is to compute an equally weighted moving average. The method is easy to program and requires only a specification of a window length. If a log is averaged by a window of 5 feet, the smoothed result will screen out variations whose thickness is less than about half of this dimensions, while retaining variation whose scale is greater. The raw log variation is convolved by a filter to a smoothed trace. A better smoothing operation would be one which caused a bestfit curve to be drawn through a data segment scaned by the window. An equivalent computation can be achieved through the design of a convolution filter which takes successive values of depth which are separated by a fixed increment on a digitized record. When a filter is applied to fit a curve through a data segment, the value of the curve is computed at the centre of the segment. The window is then moved by an incremental step and the convolution repeated to obtain the adjacent smoothed point. When all the smoothed points are linked together, the result is a continuous curve whose character is controlled by the window length and the complexity of the fitted polynomial. The values of the appropriate filter weights are therefore keyed to the requirement of the computation of the polynomial curve at the midpoint of the segment. Other filters can be designed which calculate the first derivative (slope) and second derivative (rate of change of slope) at the midpoint of the curve. In practice, the choice of polynomial filters and window sizes is a decision for the user in the extraction of components deemed to be diagnostic of systematic local changes.

The isolation of a basic motif of lithological ordering simplifies the interpretation of mechanisms of deposition and diagenesis. Where such patterns exist, they can be expected to be reflected in cyclic characteristics of wireline logs and their transforms. Signal decompositions for cyclic components is commonly resolved by Fourier analysis, which translates the variation in the time domain to a spectral representation in the frequency domain. One of the conditions of Fourier analysis is that the input data should be stationary. The subtraction of the polynomial trend from the raw data isolates the residual variation which is keyed with relatively short-term facies changes. Divisions that can be correlated over the region represent trends in either a major transgressive or regressive phase which are perturbed by minor reversals, and that the whole process has a distinctive cyclic character. Cyclic variations in lithofacies and faunal distributions within the Ordovician are most commonly attributed to eustatic changes in sea level.

In many formations, localized trends is bedding character are to be expected and this property will be reflected in a drift of frequency composition with depth. Short-term spectral methods are used to provide a three dimensional view of the frequency content of data as a function of time or depth. Ashort-term spectral, profile of a sequence can be made through the calculation of a sequence of Fourier transforms. Each transform is computed for a limited segment of data sequence defined by a window. The window is moved progressively by incremental steps, with the repetition of the transformation at each step and conversion to a power spectrum. Each power spectrum graphs the square of the amplitude of component frequencies versus their wavelengths. A composite plot of the sequence of power spectra is presented as a map of squared amplitude referenced to the two scales of time (or depth) and frequency. The size of the window is a matter for the judgement of the user. A Fouriar series represents the transformation of a signal into component frequencies matched with sine and cosine waveforms. The components are mutually independent and are calculated as a set of orthonormal trigonometric functions. Other orthonormal function sets are possible, including one proposed by Walsh (1923). Walsh function are an orthogonal set of square waves which assume the value of either one or minus one. The major advantage of the Walsh transform over Fourier transform is the great reduction in computer resources required for their calculation. If the signal takes the form of a sequence of block pulses, the Walsh transformation has a considerably simpler structure than the equivalent Fourier transform. Interbedded sequences of lithologies with sharp boundaries and relatively homogeneous character conform quite closely with a model of a complex square-wave signal. When logged by wireline tools, the discontinuities will be smoothed to some degree by the averaging process of the tool over an interval set by its vertical resolution. However, log traces still remain much of the sharp boundary features in a logged sequence.

Neutron Log— The neutron tool contains a source of high-energy neutrons which radiate spherically into the adjacent formation. These neutrons experience a progressive reduction in energy through collisions with nuclei of the atoms they encounter. The primary reduction in energy occurs in collisions with hydrogen atoms within the formation. Ultimately, each neutron is reduced in energy to a thermal state, at which point it is captured by a nucleus and capture gamma rays are emitted. One or two detectors are placed at a short distance from the source and either measure the quantity of slow neutrons or the capture gamma rays. Fig. 5. The reduction in the neutron flux is mainly a function of the hydrogen concentration within the formation which, for typical reservoir lithologies, can be equated with the pore fluid fraction. A high porosity results in a low count rate at the detector since most of the neutrons will be reduced to a thermal state and captured close to the source. Conversely, a low

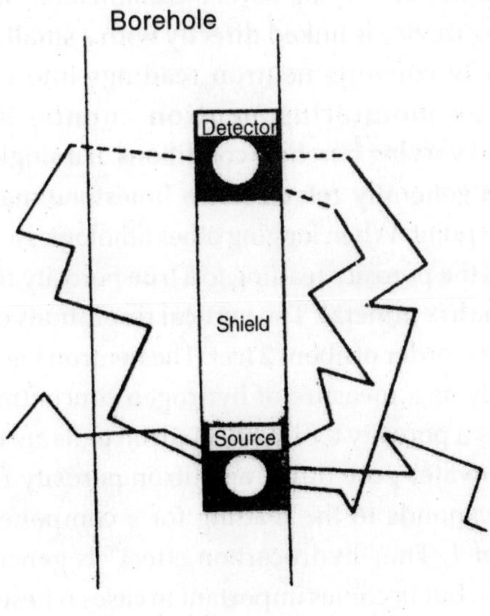

Fig. 5. Single detector neutron tool

porosity leads to a high count rate. The radius of investigation of the tool can range from several inches to several feet or metres as determined by the amount of porosity. Modern tools are scaled in either API neutron units or directly in porosity units related to an assumed limestone matrix with a pore fluid of water. There is an exponential relationship linking the porosity with the neutron reading given by the following equation:-

$$e^{-k\phi} = CN \qquad (38)$$

where ϕ = Porosity.

N = Neutron reading of the zone in counts.

k, C = Constants related to tool, borehole size, and lithology.

The modern tools are directly and arithmetically scaled in equivalent porosity units, related to either a limestone, sandstone, or dolomite matrix. The new tool mounts the source and detectors in a rubber pad which is pressed directly against the formation, thereby reducing the influence of varying borehole diameters. At the same time, the recording device is linked directly with a small computer which continuously converts neutron readings into equivalent porosity units by monitoring neutron counts, hardware characteristics, and varying borehole conditions. If a single porosity scale is used, it is generally referred to a limestone matrix as an arbitrary reference point. When logging other lithologies a correction must be applied to the porosity reading to a true porosity reading for the appropriate matrix mineral. The vertical resolutions of modern neutron logs is of the order of about 2 feet. The neutron log should be regarded primarily as a measure of hydrogen concentration, and only secondarily as a porosity tool. Now neutron tools are calibrated in terms of a freshwater pore fluid , a neutron porosity reading of 100 per cent corresponds to the reading for a component with a hydrogen index of 1. The "hydrocarbon effect" is generally very minor for most oils, but becomes important in cases whese these are appreciable gas saturations. If a zone is encountered which contains methane with a density of 0.1 g/cm^3 at the zone temperature and pressure, its hydrogen index can be estimated as 0.225. As a result, the neutron porosity readings will be reduced to values markedly below the true porosity values. By using the hydrogen index formula, apparent neutron porosities may be calculated for various hydrated minerals from their molecular compositions and atomic weights. These figures are useful when applying the neutron tool to the search for certain salts within evaporite sequences. The nautron response of shales will be variable as the quantity of hydrogen associated with the lattices of the range of possible clay minerals, proportion of associated bound water, and other factors. As a result, neutron porosity values for shales are drived from inspection of the log in the interval of interest.

Normative Log Solutions– The estimation of mineral compositions from log combinations is obviously subject to a variety of constraints. These include effects introduced by the vagaries of the borehole environment and measurement errors associated with the logging equipment. The mathematics used for solution conform to a linear model and are approximations of the nonlinear structure of reality. Mineral properties are idealized as constants rather than ranges of variation. The number of logs available is generally limited and restricts analysis to the major mineral components.

Log analysis results inevitably differ to some degree from real compositions. Petrographic and geochemical analyses of core are measurements of actual composition, although they are subject to measurement error and sample size estimate variation. Computations of mineral compositions from wireline logs are the results of a mathematical model which links physical measurements with the properties of minerals in an interpretative scheme of analysis. The aim of log analysis is to approximate reality to the closest degree possible, although exact duplication is not a feasible goal. However it is appropriate to make the conceptual distinction between a log analysis estimation of mineralogy and the actual rock composition. The normative method is a model that generates hypothetical mineral assemblages that are not unreasonable simulations of actual mineral compositions. The physical properties of rocks measured by logging tools are recast in the form of theoretical mineral assemblages. The transformation results in data that are more easily assimilated by the geologist than the raw log responses. The assemblages are computed to be a reasonable match with natural associations, but are bound by the limitations of the log analysis model. Log analysis interpretations of mineral compositions can therefore be regarded as normative solutions, whose results both compare and contrast with model estimations from core and drill cuttings. The concept has useful properties that can expand the role of logs. Normative solutions of clay mineral proportions from log analysis could be made based on a concensus on the physical properties of a standard set of clay minerals. While these estimates would deviate to varying degrees from actual compositions, they would be valuable diagnostic tools in themselves.

Overlay of Porosity Logs (Common Reference Scale)– Considerations of reservoir mineral identification and the recognition of mixed lithologies are of interest to the industrial log analyst. The computation of accurate porosities from either the sonic, density, or neutron log presupposes a knowledge of the reservoir lithology, A single log value will yield different porosity values according to the mineral selected as the zero porosity endpoint. Each porosity tool responds to separate minerals in slightly different ways, a combination of porosity logs can be used to discriminate several minerals and simultaneously solve for porosity. A variety of graphical techniques have been used successfully for this purpose and are proving to be useful geological tools in their own way.

These types of porosity log, sonic, density, and neutron are recorded in different units, microseconds per foot, grams per cubic centimetres, and neutron counts or percentage porosity units, respectively. However, the logs may be directly compared, if calibrated on a common reference scale. The most widely used scale for this purpose is defined in terms of equivalent units of limestone percentage porosity. Since this is the rnost common reference of the neutron logs, they generally require no rescaling. The density scale may be transformed by the setting the grain density 2.71 of calcite to zero porosity, the porosity fluid density for freshwater 1.00 to 100 per cent porosity, and interpolating intermediate values. By an analogous procedure the transit times of the sonic log can be converted to the limestone porosity equivalents by setting the matrix transit times of calcite and pore fluid as the two porosity extremes for interpolation. Both these operations represent simple linear transformations.

An overlay of any combination of the three porosity logs will give immediate indications of the lithology of logged units, by virtue of the different responses of matrix minerals to the individual porosity logs Fig. 6. When viewed together with the gamma-ray log, the relative disposition of the density and neutron logs identified the common sedimentary lithologies. Shales have a high neutron reading, a low density reading, and a high gamma-ray value. Limestones have a low gamma-ray value, and a coincident reutron

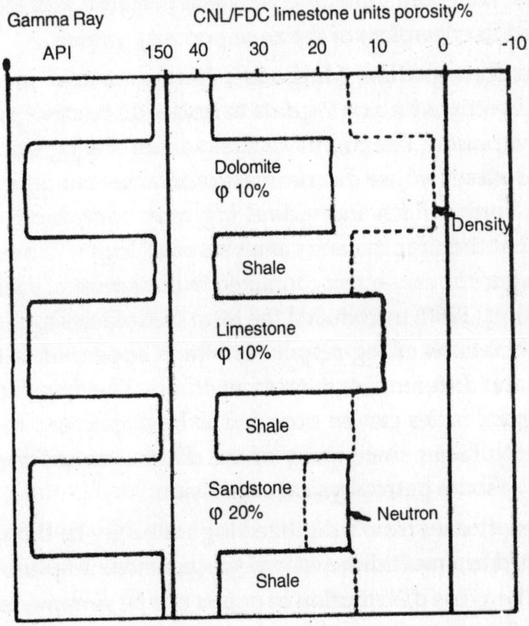

Fig. 6. Overlay of hypothetical gamma-ray, neutron, and density responses in a simple sequence of lithologies

and densitr reading, since the reference scale is calibrated to an assumed limestone. Dolomites have a low gamma-ray value, a relatively low density reading (since the grain density dolomite is higher than calcite), and a relatively high neutron reading (since the zero porosity value for dolomite is positive reading in limestone equivalent porosity). Sandstones have a low gamma-ray value, a relatively high density reading (the grain density of quartz is less than that of calcite), and a relatively low neutron reading (the zero porosity value for sandstone is a negative reading in limestone equivalent porosity). By aligning the logs on a common calcite value (zero porosity on the neutron, calcite grain density on the density, and calcite transit time on the sonic log), lithologlie character becomes apparent if appropriate allowance is made for the different degrees of stretch that should be applied to the raw unit scales. The procedure is limited to a basic recognition of grass lithological variations. Composite mixtures such as limy dolomites, dolomitic limestones, and cherty limestones are sometimes subtle to perceive, and are more easily resolved in crossplots of the zone porosity values.

Pattern Recognition Methods : Mathematical methods may be used directly with raw log data to isolate distinctive patterns of geological variation. The cumulative responses of a log suite are a multivariate data set whose discrimination of facies can take subtle and complex forms. Each individual log may contribute useful information, but the simultaneous analysis of all logs will bring out additional diagnostic characters contained in their mutual variation. Serra and Abbott (1980) introduced the term eletrofacies to describe collective associations of log responses which appeared to typify certain zones and differentiate them from others. The disposition of sedimentary rock facies can be expected to be duplicated to some degree by electrofacies since many of the diagnostic sedimentary properties have some petrophysical expression.

Readings of zones from a digitized log suite may be thought of as points located in a multidimensional space, whose reference axes are the logs. The gross distribution of points can be summarized by some basic statistics. A vector of the mean values of each log are the coordinates of the centre of the data cloud. A matrix of the variances of each log and the covariances of each pair of logs expresses both the relative inflation of the cloud along each reference axis and the dispersion in the space between the axes. Obviously, these relatively simple descriptors will not contain all the nuances of what will generally be a complex cloud pattern of points. Instead, they model the cloud as a hyperellipsoid, with its centroid located by the log means and cross-sectional profiles specified by the variances and covariances. The correlation matrix is described geometrically by an ellipse. The eigenvectors of the correlation matrix locate the major and minor axis of the ellipse. The eigenvalue associated with each eigenvector gives the relative length associated with each axis. These eigenvector are called the principal components of the correlation matrix and define a related set of axes. The proportion of the total variation satisfied by any principal component is given by its eigenvalue divided by the sum of all the eigenvalues. For standardized data, this eigenvalue sum is equal to the number of logs used in the analysis. The condensation of the readings from a large suite of logs to a smaller set of composite variables simplified considerably the task of electrofacies recognition. Zones with similar gross petrophysical aspect will tend to be clustered as localized cloud concentration of data points. The clouds will have different shapes and sizes, as dictated by the internal variability of each electrofacies and the effects of log errors. The geological meaning of each electrofacies can be deduced from cored sections of a well. Zones assigned to a common electrofacies are compared with depth equivalent cores samples to isolate descriptive themes of mineralogy and texture. If successful, this method can be extended to the prediction of equivalent lithofacies in uncored intervals. The basic idea of this approach has great potential as a means to integrate logging data with petrographic observation in sedimentological interpretations of log suites.

Log response associations with lithofacies observed in cored intervals may be analyzed directly by discriminant function analysis. In this case, lithofacies are identified a priori on the core and logged zones allocated between them. A classification function can be developed that attempts to distinguish separate lithofacies, based on log responses of their contained zones. If the function appears to be affective, then zones in intervals which are not cored can be assigned to one or other of the lithofacies on the basis of their log characters. Discriminant function analysis is both a means to distinguish samples from predefined populations in terms of measurement variables and a classification tool to type individuals of unknown affinities. The samplest application is directed to the linear discrimination of two populations. Implementation of the method is prefaced by the assignation of a test group of samples into two classes on the basis prior geological knowledge. If *k* variables are measured for each sample individual, the samples may be plotted as two data point clouds in *k*-dimensional space. The location and shapes of the clouds are specified by their multivariate means, variances, and covariances. The equation is given below :

$$Z = d_1x_1 + d_2x_2 \ldots\ldots\ldots\ldots + d_kx_k \qquad \ldots\ldots\ldots\ldots (39)$$

where d_1 is the discriminant coefficient associated with the variable x_1, and Z is the discriminant score which locates the position of any individual on the axis according to its variable measurements. Substitution of the midpoint between the sample means locates a discriminant index value. The index defines a partition point on the discriminant function which marks a boundary between the two classes. Discriminant scores may be used to classify individuals with one or other of the classes, depending on which side of the discriminant index they lie. Many number of classes can be treated through the use of canonical analysis or multiple discriminants. The mathematics of canonianalysis is a multivariate extension of discriminant function analysis. Multivariate means, variances, and covariances are computed for variables measured or individuals assigned to each of several classes. The axes are ordered in their effectiveness in total discrimination. Each has a descriptive equation which allows the computation of score of any individual set of observations as a projection onto the axis. In this case, a suite of discriminant score logs will be the final result of computation. Each log is keyed to different aspects of class distinction whose geological meaning can be interpreted from consideration of the variable loadings on its associated axis and the use of crossplots.

Porosity Log (Crossplots)– Crossplots may be made of the porosity log reading of zones recorded in their raw units as simple graphs, whose vertical and horizontal axes correspond to two porosity log scales. Since the theoretical values of pure sandstones, limestones, and dolomites can be computed for any range of porosity, these trends can be drawn on the appropriate crossplot graph to act as boundary endmembers. The location of zone coordinates with respect to these endmembers can be used to indentify that zone compositions, both in terms of mineralogy and true porosity, either qualitatively or quantitatively by interpolation. The location of a large number of zones on porosity crossplots is done by simple computer programs. On a line printer, many crossplots would show an undistinguished cloud of points, each of which would represent a variable number of similar zones. The number of zones which occur at each point is generally printed. This type of crossplot shows the relative densities of zones across the graph field. It is known as a frequency crossplot. Frequency crossplots are easiest to interpret for shale-free sedimentary lithologies and minerlas. If appreciate proportions of shale occur in some or all of the zones, allowances should be made for the introduction of an additional shale. There are no distinctive properties for shale that allow it to be located as a unique reference point on the crossplot graph. A shale point must be defined for the analytical section that best typifies the porosity readings of shale zones. This may be done from observations taken from the raw logs themselves, or by using the variation of the crossplotted zones to define its location.

The Z–plot is a useful variant of the basic crossplot that allows a third variable Z to be added to the two variables which correspond to the two axes of the graph. The most common choice for Z is a shale indicator, usually from the gamma-ray log. For any zone, the shale content is translated into an integer scale, while ranges from zero for clean zones to for shale zone. Crossplotting is done in the normal manner, but the symbol printed is the appropriate Z value. The resulting cross-plot is a two- dimensional mapping of a three-dimensional situation in which the plotted values represent relative heights above the plane, and the vertical axis is the Z variable.

The M–N plot is directed to the task of identifying the mineral unknown in the zones and estimating their relative proportions. Composite variables, M and N, may be computed, which, by taking pairs of porosity values at a time, eliminate the contribution made by the pore volume component. The computation of M and N values fòr a variety of matrix minerals are computed prior to log analysis and located on a graph of M versus N. M and N values express the slopes of hypothetical lines linking the mineral with a porosity fluid component. Fig.7. Having located the thoretical positions of an appropriate suite of minerals on the M-N plot to act as reference points, M and N values may be calculated from each zone porosity reading and located on the plot. The disposition of the zones gives immediate indications of their matrix mineralogies. For any given zone, the composition may be resolved in terms of three minerals as

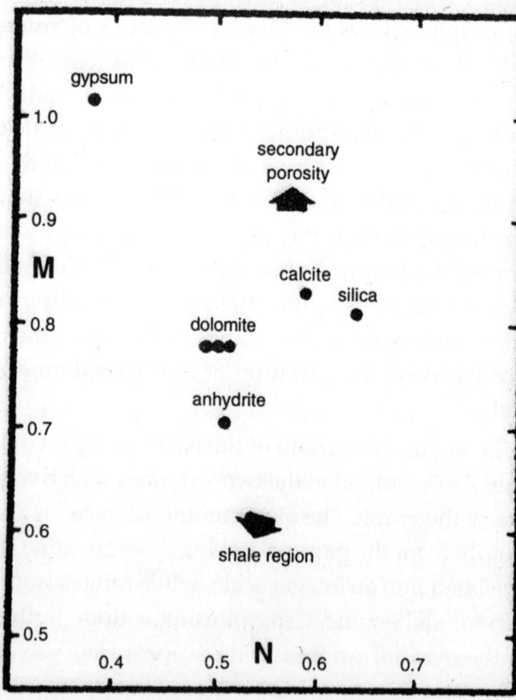

Fig.7. Simplified M-N plot, showing locations of common minerals.

an upper limit to the number of unknowns that may be accommodated uniquely. In zones which have appreciable residual gas saturations expressively large values of M and N will result, and so will be immediately apparent. The MID-plot technique isolates apparent density and apparent sonic travel time as parameters which are independent of porosity. On a crossplot of bulk density and neutron porosity apparent density isoliths are located by interpolation between sandstone, limestone, and dolomite lines and extraploation beyond their limits. On the three fundamental lithology lines, the apparent densities have real meaning as actual densities of the matrix minerals quartz, calcite, and dolomite. For other minerals, the apparent densities are artificial quantities, but are not grossly dissimilar from actual values. The same methodology is applied in the contouring of apparent sonic travel times on a cross plot of sonic travel time versus neutron porosity. The apparent densities and transit times for any logged zone can be estimated from interpolation on available MID-plot charts. How lithodensity tool is used for the measurement of the photoelectric absorption cross-section index P_e which shows excellent discrimination between minerals. The index is determined by the number of electrons per atom, which corresponds to the atomic numbers. When the index is multiplied by the electron density, the resultant photoelectric macro-section U is related volumetrically the mineral composition, with minimal influence by pore fluids. For any zone in a stratigraphic section an apparent matrix photoelectric macrosection (UMA) may be calculated, together with the apparent grain density (ROMA) from bulk density and neutron porosity readings. Location of these coordinates on a ROMA-UMA lithology crossplot allows identification and volumetric estimation of mineral species, using concepts similar to the M-N and MID plots.

Grossplots can be considered as simply pictorial presentation of log data. The properties of projected geometry allow grossplot techniques to be devised as problem-solving procedures. In any given subsurface study, appropriate logs may be selected and their combinations projected in a manner which best accentuates or discriminates trends and patterns which are of interest to the analyst.

In the analysis of log data, grossplots are a natural precursor to numerical analysis of compositional variation by an algebraic methods. This preliminary phase is necessary to relate the complexity of surface variations to the basic framework of a descriptive model with geologically meaningful properties. The simple frequency crossplots of raw logs represent orthogonal projections of data variation. M-N and MID plots are oblique projections in three dimensions. Rather than view the data clouds from a hypothetical infinite viewpoint, they are conical projections in which the observer's eye is moved to the fluid point. The fluid point is a useful focus, since the projection causes an effective suppression of porosity variation and so is particularly useful in discriminating mineralogy changes in the common reservoir rocks. Alternative projections and focal points can be chosen to meet the requirements of different types of studies.

Porosity Trends and Patterns (Geological Implications)—The widespread use of the scanning electron microscope has literally provided a new insight into aspects of the pore phase which are often obscured in conventional thin section. The pore network is more clearly perceived in terms of depositional fabric geometry, function as a conduit to modifying solutions, and host to diagenetic minerals. Recent studies have attempted to unravel petrography as a multiple record of deposition, tectonism, hydrodynamic, and geochemical regimes, which are basic historical elements of basin development and stratigraphy. Since porosity provide the link between the fluid and solid phases, sedimentary formations should be viewed in terms of phase differentation. This theory may be useful unifying concept for the processes of mechanical compaction and diagenesis.

Early work on porosity variations in sandstones tended to emphasize the role of the fabric as an inert pack of quartz grains subjected to compactional rearrangement, with a degree of pressure solution and redistribution of dissolved silica. In experiments with unconsolidated sands showed positive relationship between porosity and textural parameters of sorting, low sphericity, and angularity. Diagenesis is not restricted to the destruction of primary porosity, but ultimately enhances pore volumes in the creation of significant secondary porosity. A systematic inverse relationship was found between grain size and porosity in carbonate rocks, but a negative correlation was found between porosity and permeability in contrast with carbonate rocks. Porosity developments in ancient limestones are the cumulative result of a diagenetic sequence which may involve multiple histories of cementation, dissolution, and mineralogy changes. In cases where the original sediment framework showed marked differentiation between contiguous depositional facies the characteristic may be presented as distinctive porosity zonations on logs.

Processes of compaction and chemical diagenesis cause a wide variation in pore volumes and porosity types within carbonate rocks. The transformation of calcite to dolomite involves a contraction on the order of 12 percent, with associated increase in porosity. Although porosity changes are often observed, molecule-for-molecule replacement is only one aspect of the complex processes of dolomitization. Dolomitization and pore development often appear to be strongly fabric selective, and result in a concordant diagenetic overprint of contrasts in depositional facies. A direct relationship was found between the extent of dolomitization and the amount of micrite associated with crinoid fragments within the original sediment. Crinoidal limestones with rim cement surrounding the crinoids showed partial occlusion of pore space. Geological studies of the interrelationships between porosity development, depositional facies, and diagenetic modications are becoming increasingly common in the literature. Their conclusion are of great importance in the interpretation of porosity trends shown by sonic-neutron, and density logs, as well as resistivity traces in water-saturated sections. Conventional sources of geological information provide the necessary key to both initiate and examine alternative genetic models, e.g., core, cuttings. But this kind of data is generally restricted both areally and vertically when compared with the extensive coverage supplied by logging suites. The intelligent coordination of both types of data extends the use of porosity logs beyond their traditional roles as stratigraphic correlation traces and evaluators of potential reservoir units.

Remedial Corrections of Logs—All physical measurements are subject to error. Systematic errors are introduced by environmental effects of boreholes rugosity, variable mudcake thickness, and degree of formation invasion by mud filtrate. Even after the most comprehensive environmental corrections, the log trace still includes a distinctive source of error which is compounded from tool malfunctions, shop and field miscalibrations, and operator error. Procedures that attempt to eliminate logging measurement error are usually referred to as normalization or quality control. The most commonly used methods are given below:—

1. **Use of Calibration Units**—The well bore may penetrate certain stratigraphic horizons that can be used for internal calibration of the well logs. Certain shales or relatively tight carbonate horizons are often selected on the basis that they can be correlated over a wide area and their physical properties appear to vary only to a minor degree. All lithologies show some degree of varietion in composition and texture. But there will be a narrowly restricted range of variation. Changes in calibration units properties will be linked with geography which can be corrected.

2. **Application of Crossplots**—Sub-surface units with simple mineralogies, e.g., clean sandstone, pure limestones, and dolomites, and logged by two or three porosity logs can be crossplotted and their trends compared with theoratical sandstone, limestone, and dolomite lines to determine the necessary shifts to be applied to the logs. M-N plots may also be used for this purpose. Resistivity logs can be corrected through computations of theoratical porosities in water saturated units and comparison with porosity logs on a dual porosity crossplot.

3. **Comparison with Core Analyses**—If cores are taken from a interval within the well, engineering data of porosity and grain density may be used in the correction of the porosity logs. A comparison of the mean porosities from core and porosity log allow the computation of the necessary shift to be applied to the log. The various logs which are sensitive to porosity must be shifted to common depth registration with one another and with the core interval. Systematic deviations between core and log data would form the basis for log normalization.

4. **Overlays**—In neighbouring wells with similar stratigraphy, an overlay of the same log from two wells, correlated to match equivalent units gives an immediate indication of the relative shift between their separate calibrations. This method does not directly indicate whether one or both of the logs requires correction, or indeed the magnitude of the corrections.

5. **Histograms**—If the same log in different wells is digitized for a stratigraphically equivalent interval, histograms of the relative frequency of different readings may be compared to determine a relative shift to match the histogram peaks and troughs in moving between the wells. Some external criteria are needed to provide explicit information on the necessary corrections.

6. **Simple Mapping**—The average log response in a correlatable interval from a set of wells can be mapped areally as a continuous surface. Localized anomalies in the resulting contour map indicate possible problem wells whose calibration appears to deviate significantly from the surrounding well control. This method is most successful when applied to an interval whose log response is likely to show only gentle regional changes.

7. **Trend Surface Analysis**—The empirical method of simple mapping may be formalized in a statistical model through the use of trend surface analysis. The theory is drawn directly from regression, which partitions observational data between a regional systematic component and a spatially uncorrelated random element. The regional component is estimated by a regression of the mapped variable of polynomials of the geographic coordinates of the well control. The regional surface is fitted to the data in such a way that the sum of the squares of their deviations from the surface is the minimum possible. Fig.8. A hierarchy of trend surfaces matches polynomial equations of differing degrees of complexity. A linear surface is a plane, a quadratic surface is a paraboloid which may describe a basin or a positive feature. Higher order trend equations will represent successively more complex surfaces. As a regression procedure, trend surface analysis functions as an analysis of variance, in which the total variation of the data is partitioned between variation accounted for by the surface

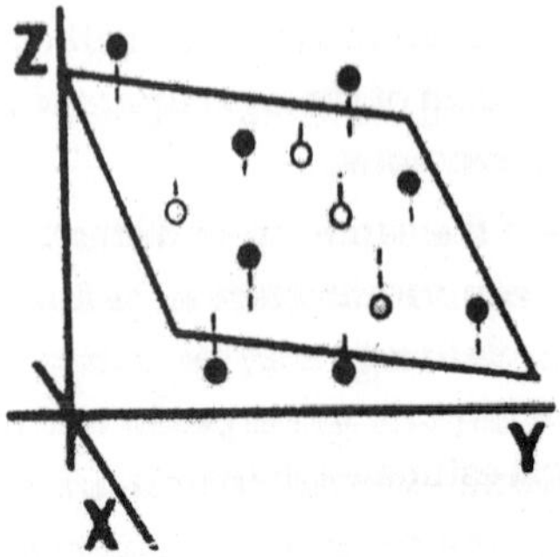

Fig.8. Hypothetical linear trend surface fitted to a variable, Z, measured at locations with cordinates X and Y.

and the remainder as the sum of the squared deviations from the surface. When the residuals show no spatial correlation, the associated surface contains an estimate of the systematic spatial structure of the data variation. This model can be applied to the mapping of a log response in a calibration unit. A calibration unit is defined as a geological horizon which is judged to show only minor variation across the region of interest. A log response measured at any location can be subdivided into two components : (1) a systematic regional element, and (2) a random error associated with the miscalibration of the tool. The trend surface normalization method can be expanded to a generalized strategy of analysis designed for routine application. For more detailed studies, analyze the residual distributions in assessments of error character contrasted between tools and between service companies in terms of relative precision and bias. Update calibration unit data sets with log responses and geographic coordinates from newer control as it becomes available. Periodically, recompute trend surfaces as successively more refined estimates of regional variation.

Resistivity Measurement—All resistivity tools are to some extent "averaging" devices that record resistivities of zones rather than resistivities of discrete points. The drilling mud forms a mudcake seal on the borehole wall of permeable formations. In doing this, mud filterate penetrates the formation, displacing formation water and oil or gas. In a zone immediately adjacent to the borehole the mud filterate displaces all the formation water and any movable oil saturation. Beyond this, the mud filterate displaces part of the formation water in a "transition zone" which ultimately peters out at a contact edge with virgin formation. This process is termed "invasion" and the relative depth of invasion is broadly a function of formation porosity/ permeability properties. A resistivity tool is selected that will best estimate the resistivity of the virgin formation by taking into account borehole characteristics, drill mud properties, formation lithologies, and degrees of invasion. Four classes of the most common resistivity devices are given below :

1. **Normal Device**—The normal device tool consists of two electrodes separated by a fixed spacing on an insulating mandrel

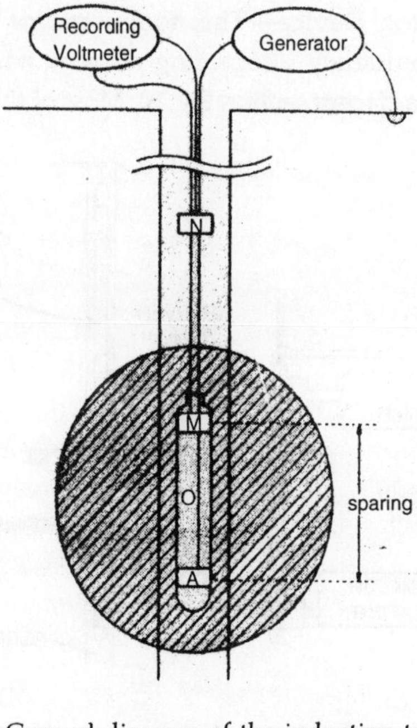

Fig. 9. General diagram of the induction tool.

Fig. 9. A current is passed between the lower electrode and another electrode at ground level and the potential is measured at the upper electrode and an electrode intermediate between it and ground level. The current and measured potentials allow the computation of the resistivity of a zone whose reference point is midway between the two tool electrodes. The resistivity corresponds to that of a volume surrounding the tool whose radius is approximately twice the tool electrode spacing under favourable conditions. As a result the measurement tends to be a composite of resistivities from the mud, invaded zone, and virgin formation.

Since the "radius of investigation" is largely dictated by the electrode spacing, two common variants of the device are the "short normal" (16-inch spacing) and the "long normal" (34-inch spacing). Now this device is not much in use. Sometime the short normal is run with the induction resistivity tools and comparison of its trace with the induction curve gives indications of relative degrees of invasion.

2. **Induction Device**—This tool contains transmitter coils with a high frequency alternating current which induce eddy currents in the adjacent section Fig.10. Most of these eddy currents

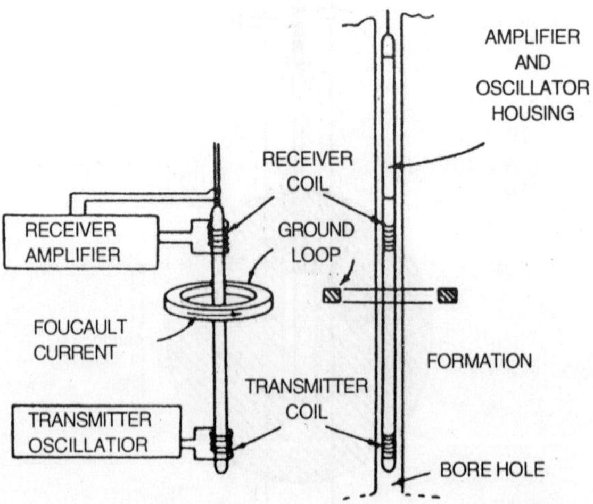

Fig.10. Detail diagram of the induction tool.

are focused beyond the diameter of the typical flushed zone and their magnitude is an approximation of the conductivity of the virgin formation. In turn, they induce voltages in the receiver coil which are transformed to estimates of formation conductivity and, as reciprocal resistivity.

The induction tool is ideally suited for sandstone sections, which typically have high porosities, but may not be a satisfactory choice in highly resistive sequences such as low porosity carbonates. The induction tool can be run in holes drilled with air or with oil base muds since it does not require electrical contact with the mud or formation. The tool operates well in "fresh mud" but readings become strongly degraded in "salt muds".

3. **Laterolog**— The laterolog provides accurate readings of formation resistivity in holes drilled with saltwater mud. The central principle of operation is a three electrode arrangement in which a current of constant intensity is supplied to the central electrode. Fig.11. A variable current intensity is transmitted to the two

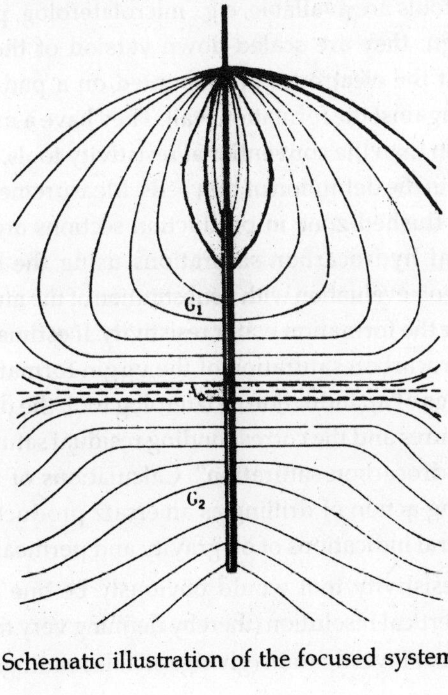

Fig.11. Schematic illustration of the focused system lateslog tool.

surrounded electrodes whose magnitude is adjusted so that there is a zero potential with the central electrode. As a result, current in the central electrode is constrained to flow radially outward as a "current disc" into the surrounding formation. The thickness of the disc is determined by the spacing of the surrounding or guard electrodes while the current density at any lateral distance from the central electrode is inversely proportional to this distance times the spacing. The drop in potential of the current disc radiating into the formation is monitored by a remote return electrode. As a result, an apparent resistivity is deduced which is the sum of resistivity contributed by the mud, invaded zone, and virgin formation. In situations where the mud is relatively conductive, degree of invasion is restricted and resistivity of the formation is fairly high. This apparent reading is a close approximation of the true virgin formation resistivity.

4. **Microresistivity Tools** : The function of micro resistivity tool is to measure the resistivity of the flushed zone. A variety of microresistivity tools are available, e.g., microlaterolog, proximity log, etc. In design, they are scaled-down version of their larger counterparts, but the electrodes are mounted on a pad which is pressed directly against the formation wall. They have a much finer vertical resolution than the conventional resistivity tools. They are especially useful in the definition of thin beds. Measurements of the resistivity of the flushed zone in production sections are used to calculate residual hydrocarbon saturations using the standard formulas of reservoir evaluation with a substitution of the mud filtrate resistivity, R_{mf} for the formation water resistivity. If estimations are made of the hydrocarbon saturation of the virgin formation from resistivity readings of an induction or laterolog tool, the difference between these figures and the corresponding residual saturations is the "movable hydrocarbon saturation". Calculations of this type utilize the flushing action of drilling as an ersatz production tests which gives general indications of oil gravity and permeability.

The ideal resistivity tool would obviously be one that had extremely good vertical resolution (thereby defining very thin beds) and a large radius of investigation (giving reliable readings deep in the formation). These two criteria are almost impossible to meet in the design of a practical resistivity device. As a result, different resistivity tools are run for different purposes and several types may be run in the same hole. Analysis by hand is done most rationally after zoning the logs as an initial step. The vertical resolution of the induction and larger laterologs ranges between 3 and 5 feet.

Resistivities of Rocks—A resistivity log from a borehole represents a composite measurement of resistivities of rock-forming minerals and fluid and gaseous phases within the pore spaces. Most sedimentary sequences consist almost entirely of sandstones, limestones, dolomites, shales, and cherts, with sporadic occurrence of evaporites and coal beds. The resistivities of their component minerals are all in excess of a million ohm-metres. Similar high resistivities also characterize hydrocarbon oils and gases. The effective measurement scale of resistivity logging tools is generally limited to a range between 0.1 and 2000 ohm-metres. The resistivity log is obviously controlled by factors other than the resistivities of mineral components. Shales have low resistivities which are caused by clay mineral properties. In sandstone, limestone, and dolomite beds which are shale free, electrical current is almost entirely carried by the aqueous phase of the rook pore space. Pore water can generally be considered as a simple salt solution. The resistivity of a subsurface brine varies inversely with the concentration of dissolved solids. Brine resistivity decreases with increasing temperature. Resistivities of formation water tend to decrease with depth in the sub surface as a result of increasing concentrations of dissolved solids and increasing temperatures. In a unit that contains oil and gas, high resistivities match low porosities and low resistivities indicate higher porosities. Since electrical current is conducted by the brine contained in the pore space of a shale-free rock.

The relation between formation factor F and porosity is ϕ given by the following equation :

$$F = \frac{1}{\phi^m} \qquad \text{............ (40)}$$

Where *m* is some value greater than unity which is called cementation factor. The quantity m has a value of about 1.3 for unconsolidated sands and has a range between 1.8 and 2.0 for many consolidated sandstones. In fact, *m* is regulated by the tortuosity of the pore network, but this is the complement of the matrix geometry which is in turn controlled by the rock texture. The famous "Humble equation" is given as :

$$F = \frac{0.62}{\phi^{2.15}} \qquad \text{.............(41)}$$

The textural interpretation of the cementation factor can be drawn from the analysis of a subsurface sandstone logged by a resistivity and a porosity device. Interpretation must take account of both the textural properties of the original sediment and the degree to which the primary characters may have been obscured in compaction and diagenesis. Field studies should be coordinated with explicit petrographic and sedimentological information from available core and cuttings for genetic interpretations. Cementation factors measured from carbonate core samples show generalized associations with textural character. The use of cementation factor profiles as a means to characterize carbonate framework properties is more speculation than is the case for sandstones. In practice, meaningful result would require particularly close integration of core data on an individual case study basis to relate pore morphology suggested by resistivity with more conventional genetic descriptions. The log characterization of textural properties in oil and gas reservoirs is complicated by the hydrocarbon phase as a new source of variability. However, systematic changes in matrix geometry are reflected in the fluctuations of irreducible water saturation with porosity. Plots of irreducible water saturation versus porosity in relatively homogeneous reservoirs generally form a distinctive trend which is approximated by a hyperbolic curve.

Shale Compactional Trends (Analysis)– Shale units are recognized easily on the sonic, density and neutron logs. Shale zones generally register high porosity values which are a response to water content rather than the effective porosity of potential reservoir lithologies. Water within the shales is the major control on variation in the physical properties of density, bulk modulus of elasticity, and hydrogen concentration. The properties show a broadly monotonic change with depth which reflects compactional effects as the result of mechanical, chemical and thermal mechanisms. The degree of compaction is matched by decreasing transit times, increasing densities, and decreasing neutron porosity readings. Fig.12 shows the variation of average Gulf Coast shale bulk densities with depth.

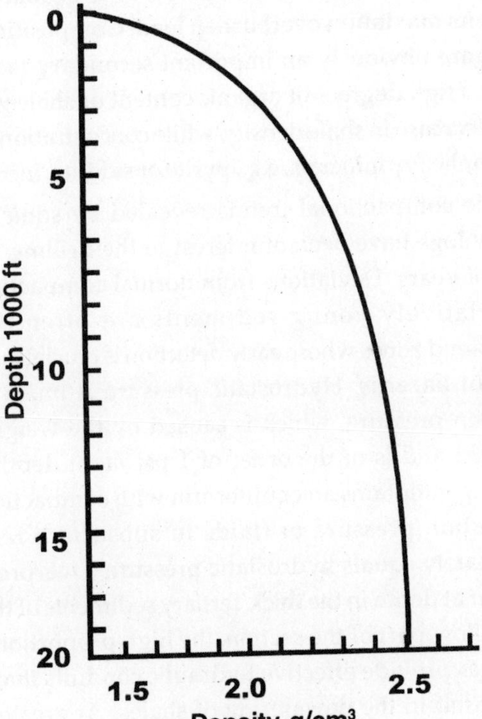

Fig.12. Variation of average Gulf Coast shale bulk densities with depth (After Eaton, 1969).

with depth. The essential process is one of sediment dewatering. Diagenetic changes in easy mineralogy are crucial element of more realistic compaction models. Geothermal controls are considered to be more influential than overburden pressure. Alterations is clay mineralogy, primarily smectite to ellite, are a substantial component of dewatering and occur at several discrete stages of temperature and pressure. The dewatering fails to keep pace with hydrostatic pressure, overpressured zone can develop. In cases of marked differential loading, these is upward diapiric movement of shales, and even penetration of the surface as mud lumps. If the degree of compaction could be related to the depth of burial, it would be possible to estimate the maximum depth to which a shale had been buried. A shale could be considered as a geobarometer which recorded its maximum overburden load. Compositional variations of shales are obviously an important secondary factor in density variation. High degrees of organic content or shale gas can cause a marked decrease in shale density, while concentrations of calcareous material or heavy minerals, e.g., pyrite or siderite, increase densities.

Shale compactional trends revealed by sonic, density, and resistivity logs have been of interest to the drilling industry for a number of years. Deviations from normal compactional trends in thick, relatively young sediments are often diagnostic of overpressured zones whose early detection is crucial in the avoidance of blowout hazards. Hydrostatic pressure is much less than the overburden pressure, which is caused by the weight of the rock overburden and is of the order of 1 psi /foot depth. If sediment dewatering maintains an equilibrium with compactional gradient, the formation pressure of fluids in subsurface reservoir units approximately equals hydrostatic pressure. Overpressured zones often occur at depth in the thick Tertiary sediments of the Gulf Coast. In the shallow part of the section, the high proportion of coalesced sand-stones provide effective hydraulic conduits that transmit the fluid released in the dewatering of shales. At greater depths, the sedimentary facies consists of deepwater marine shales with thin sandstone bodies. Subsidence and compaction may proceed at a rate that exceeds the capacity of the sandstones to remove fluid, particularly if the sandstones are isolated hydraulically. Consequently, excessive pressures may be genesated within the sandstones while the adjacent shales are characterized by high fluid pressures. Shales in the overpressured zone can be recognized as undercompacted relative to their expected degree of compaction on a normal compaction depth trend topical of sequences in which shale dewatering has kept pace with overburden stress. The sonic log has been widely used in the detection of overpressured zones.

The detection of overpressured zones by logging techniques has immediate importance in the engineering of exploration wells in thick sequences of relatively young sediments. At the same time, it gives important insights into the nature of the compaction mechanism as a current dynamic process. The integration of geochemical, geothermal and hydrodynamic theories with earlier empirical methods is likely to refine conclusions from studies of compactional trends in ancient basins.

Shaly Sandstones (Graphical Methods)-- The analysis of sequences with significant contents of shale is made complex by the wide range possible for clay mineral composition, nature and volumetric content of the silt-sized fraction, and degree of compaction. The log analysis of shales and shaly intervals becomes a two-stage procedure. In the initial phase, shale properties which typify the studied section are identified from the logs themselves. On a second pass through the data these representative endmember values are used to calculate volumetric estimates of shale content. Grossplots are invaluable in the reconnaissance phase. Their graphic nature is an ideal medium for the pattern recognition skills of the log analyst and geologist. The variability of shales within even short sections requires intelligent decisions concerning what constitutes representative shale properties. The crossplots contain information on trends of variation which are diagnostic of fabric changes and useful in interpretation of sedimentation and diagenesis patterns.

Shaliness affects the various log responses in terms of its volumetric proportion, physical properties, and the manner that it is distributed in the rock. Three possible endmembers are laminar, structural, and dispersed shale. Laminar shale units consist of thin laminations of shale which separate stringers or beds of sandstone. There is an overall reduction of the bulk porosity of the total rock. Structural shale describes sandstones in which some of the grains of the framework are shale fragments or clay mineral diagenetic alteration products. Dispersed shale consists of clay minerals which develop in grain framework surfaces and occlude pore space. Extensive development of a dispersed shale content leads to a progressive reduction in porosity. Fig.13. It is relatively easy to assign different categories of authigenic and allogenic clays to the shale distribution types. Fig.14. A variety of indicators are used from various logs in the assessment of the proportion of shale in a zone. The more commonly used shale indicators are the gamma-ray log, SP log, and neutron density log combination. In sections which have been logged by a variety of tools, separate estimates of the shale content are computed for this shale indicators which are available. The lowest estimate is usually to taken to be the most valid.

A neutron-density log combination is often used as an alternative method to evaluate shale content of shaly sandstones. Formation evaluation is complicated by the extremely low-formation water salinities which often vary dramatically between sandstone

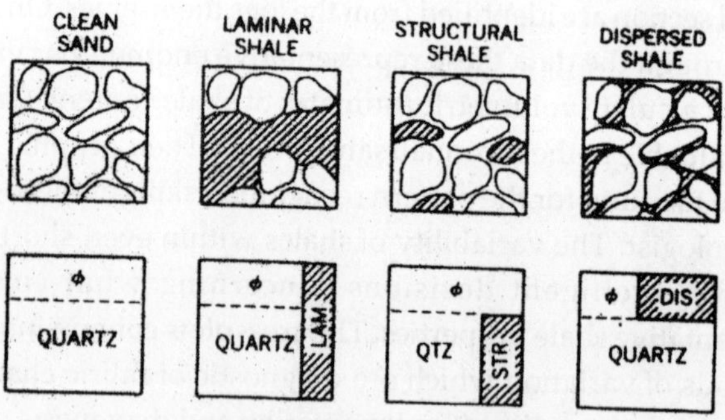

Fig.13. Distribution types of shale in shaly sandstones as hypothetical limiting cases of a simple model for log analysis.

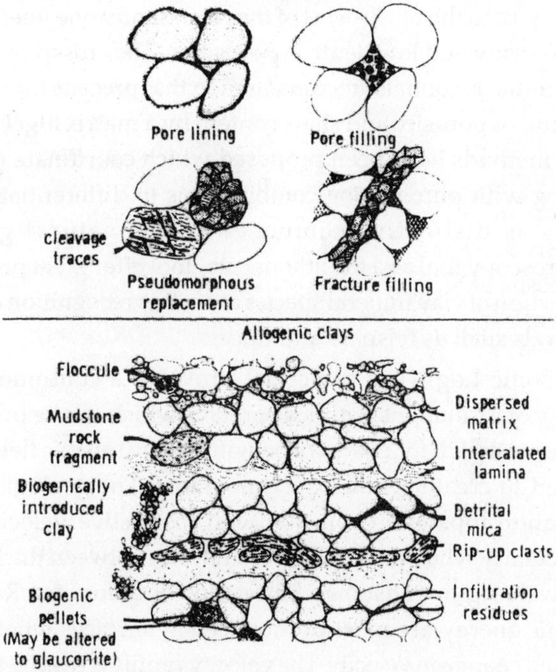

Fig.14. Modes of occurrence of authigenic and allogenic clays in sandstones (After Wilson and Pittman, 1977).

units. A cross-plot of neutron and density values for zones within the section of interest can be referenced to a compositional triangle of quartz, water, and shale. A sandstone line may be drawn, which links the log responses of quarts and water. A shale point is located at the more extreme values of the plotted zones, in such a manner as to complete the composition triangle by-straight line segments which link it to the quartz and water endpoints. Following the precepts of the log analyst model of shale composition, trendz may be drawn which correspond to structural shale (isoporosity) dispersed shale (isoquartz), and laminar shale. Referenced to these trends the disposition of the zones suggests that the shale element of the shaly sandstones is disseminated as a mixture of pore-filling clays and shale laminations. Gas zones are immediately recognized as points which plot to the north-west of the clean sandstone line, due to their high density and low neutron porosities. The crossplot can be used as an initial reconnaissance evaluation that preceds the quantitative solution of porosity and shale content by a matrix algebra solution. New methods have been proposed which coordinate the gamma-ray log with porosity log combinations to differentiate clays and micas as distinctive components. The natural gamma-ray spectroscopy tool and the lithodensity tool offer great promise in the distinction of clay mineral species as well as recognition of accessory minerals such as felspars and micas.

Sonic Log—The sonic tool provides a continuous velocity survey of formations within a borehole which can be processed as a reference match for comparison with nearly seismic field records. It is used to create synthetic seismograms. The geologist's picks of formation tops often coincide with distinctive velocity changes, particularly when these match boundaries between thick shales and relatively rigid sandstones, limestones, or dolomites. Reflections of seismic energy are most pronounced at surfaces where there is a marked change in velocity. The velocity profile is readily transformed into a sequence of reflection coefficients, which are then convolved with a wavelet whose frequency character matches that of the energy source used for field seismic work. The end result is a synthetic seismogram in which reflection events can be tagged with specific formations and used as a key to identify corresponding events on seismic records.

The potential of the sonic log for porosity evaluations was made possible by the strong contrast in sound velocity between the mineral framework of a reservoir rock and the fluids contained within its pore space. The schematic design of the sonic tools is shown in Fig.15. In old sonic tools, a transmitter emits sonic pulses at a rate of 10 Hz. The pulse frequency is centred around 20 KHz. Acoustic energy emanates from the transmitter as a compressional wave which crosses the relatively slow velocity medium of the borehole fluid. It is refracted at the borehole wall and travels through the formation at

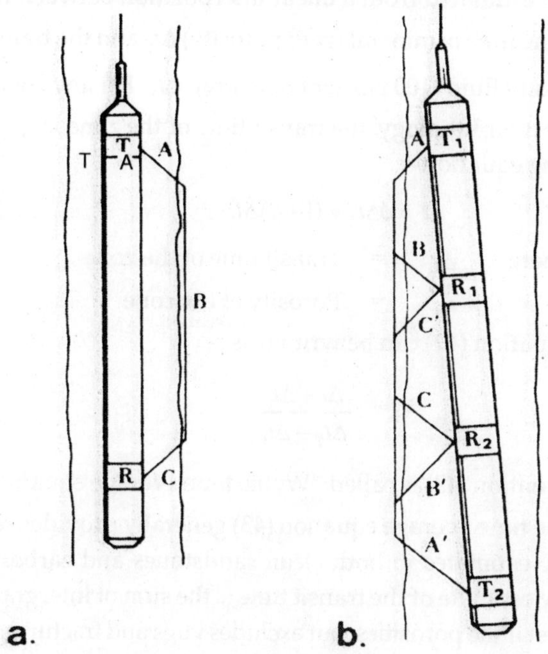

Fig.15. Diagrams of sonic tool (a) Earlier sonic tool (b) Modern borehole compensated sonic tool.

a greater speed caused by the increase in density and rigidity. Compressional waves are about twice as fast as other waveforms and are the first to reach the receiver. The receiver records the time between the emission of the signal and its reception. The old sonic tool was modified to the modern borehole compensated tool. The modern tool uses two transmitters and two receivers. The thickness of formation sampled corresponds with the distance between the two receivers. When the sonic tool is drawn upward through the borehole, its measurements are converted to a velocity profile of the sequence as a function of depth. However, sonic logs are conventionally recorded in terms of "transit time". This is the time in microseconds taken for the compressional wave to travel through one foot of formation and is the reciprocal of its velocity. Porosities could be estimated from a linear interpolation between the transit time of the matrix mineral (zero porosity) Δt_1 and the transit time of the porosity fluid (100 per cent porosity). Δt_2. For any zone of shale-free reservoir lithology, the transit time of the zone is given by the following equation–

$$\Delta t = \phi \Delta t_2 + (1 - \phi)\Delta t_1 \qquad \text{.............(42)}$$

where Δt = Transit time of the zone

ϕ = Porosity of the zone

Equation (42) can be written as :–

$$\phi = \frac{\Delta t - \Delta t_1}{\Delta t_2 - \Delta t_1} \qquad \text{..............(43)}$$

Equation (43) is called "Wyllie time average equation"

The time average equation (43) generally provides reasonable porosity estimates in both clean sandstones and carbonates. The porosity estimate of the transit time is the sum of lntergranular and intercrystalline porosities, but excludes vugs and fractures. The sonic estimate then corresponds to a primary porosity and the other tool estimate to a total porosity. The difference between the two porosities is the secondary porosity, which is an estimate of the proportion of vugs and/or fractures. In unconsolidated or semiconsolidated sands, an additional correction must be applied to porosities estimated by the time average equation. The apparent porosities of the time average equation are converted to true porosities through multiplication by a constant. This constant is generally derived from numerical comparisons of apparent porosities with porosities from either neutron, density, or resistivity logs, or from available core porosities. The matrix transit times of lithologies such as sandstones, limestone, and dolomites are defined by narrow ranges, since these correspond to values for quartz, calcite, and dolomite, which may be measured in the laboratory. The transit times of shales are drawn from the sonic log itself.

Spontaneous Potential (SP) Log—Natura1 potential are observed to occur in boreholes which are measurable in millivolts and finds to distinguish permeable formations from intervening shales. The cumulative electromotive force that is generated is comprised of a shale potential, a liquid junction potential, and an electrokinetic potential. The electrokinetic potential is very small. Shale potential and liquid junction potential result from the permeability of shale to sodium ions, and the greater mobility of chloride ions relative to sodium ions, respectively. These two types of potential are classed as the elctrochemical component of the SP log and result in an electrical circuit at the boundaries of shales with permeable beds. The flow of the current is from permeable bed to shale to borehole mud and back to permeable bed. Fig. 16. The magnitude of the electrochemical potential E is a junction of the chemical activities of the formation water and the mud filterate. If

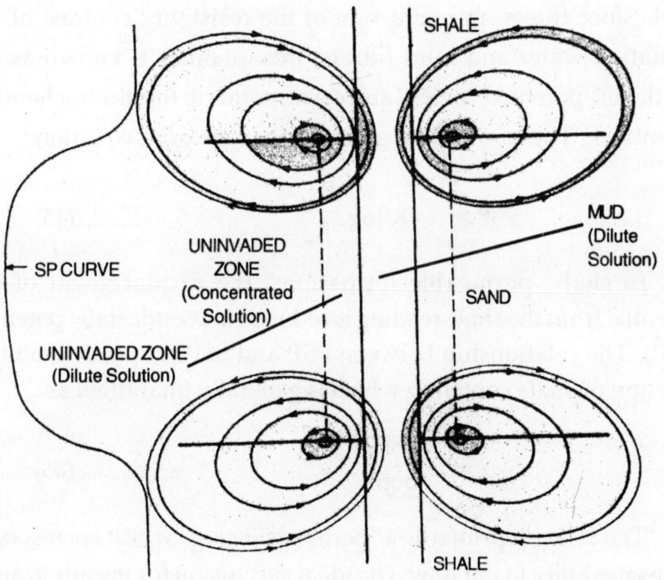

Fig.16. Electric current movement at the boundaries of permeable and impermeable beds in the vicinity of a borehole.

mud filterate and formation water can be considered as sodium chloride solution then E can be given by the equation.

$$E = -K \log \frac{R_{mf}}{R_w} \qquad \text{.............(44)}$$

where R_{mf} = Resistivity of the mud filtration

where R_w = Resistivity of the formation water.

where K = Constant

The SP log is run in conjunction with the deep induction log and short normal as an induction electrical survey. The SP defines the location of permeable formations and intervening shales and, for fresh mud boreholes, shales read as relatively positive readings and permeable formations as relatively negative readings. The measurements are made with reference to an electrode at ground level. Since this is an expression of the resistivity contrast of the formation water and mud filtrate, this quantity is known as the "static self-potential" or SSP and corresponds to the electrochemical potential E. Therefore SSP is given by the following equation :

$$SSP = -K \log \frac{R_{mf}}{R_w} \qquad \text{.............(45)}$$

In shaly, permeable formations, the displacement of the potential from the shale reading is reduced to pseudostatic potential (PSP). The relationship between PSP and SSP is approximately a function of shale content α which is generally quantified as

$$\alpha = \frac{PSP}{SSP} \qquad \text{.............(46)}$$

The SP log is primarily a "permeability log" which is a response to permeability to ion flow. The ideal succession for the application of the SP log is a sandstone-shale sequence comprised of relatively thick beds. The SP log has variable success in the definition of limestone and dolomite beds in carbonate shale sequence. At relatively tight zones or thin carbonates, the trace often shows little differentiation from the shale baseline. The reduction in SP response opposite oil and gas zones in reservoir units has been noted. The presence of oil or gas shifts the SP trace towards the shale baseline. The effect is greatest at higher hydrocarbon saturations and is more pronounced in shaly sandstones than in relatively shale-free equivalents.

The SP response of a sandstone provides a profile whose shape gives general indications of the nature of the upper and lower contacts, degree of interbedding and gross grain-size variations. Sharp contacts give abrupt changes in potential at shale sandstone boundaries. The occurrence of alternating beds of shale, shaly sandstone, and sandstone within the main sandstone unit will be marked by a serrated character to the SP trace, provided that those beds are not too thin to be resolved individually by the tool. The possible variations in SP shape for thick sandstone bodies are shown in Fig. 17. The SP profiles of subsurface sandstones may be interpreted in terms of geometry and texture, with subsequent classification of depositional environment. The technique provides a useful method of extrapolate interpretation of sedimentary facies of units into the subsurface, using profiles which can be related directly to sedimentation models. In the analysis of facies variations, the SP trace is usually coupled with a resistivity profile as a dual curve motif. In general, the resistivity log forms an approximate mirror image of the SP, although it often shows more detailed character, since it is sensitive to porosity changes, unlike the SP. SP logs are far more common than cored sequences and often provide detailed information on the shape of subsurface units expressed either as cross-sections or onmaps. They give clearer visual impressions of vertical variability than is available from coarsely sampled drilling cuttings. Obviously, their intelligent application is as a contributory facet within an integrated geological interpretation.

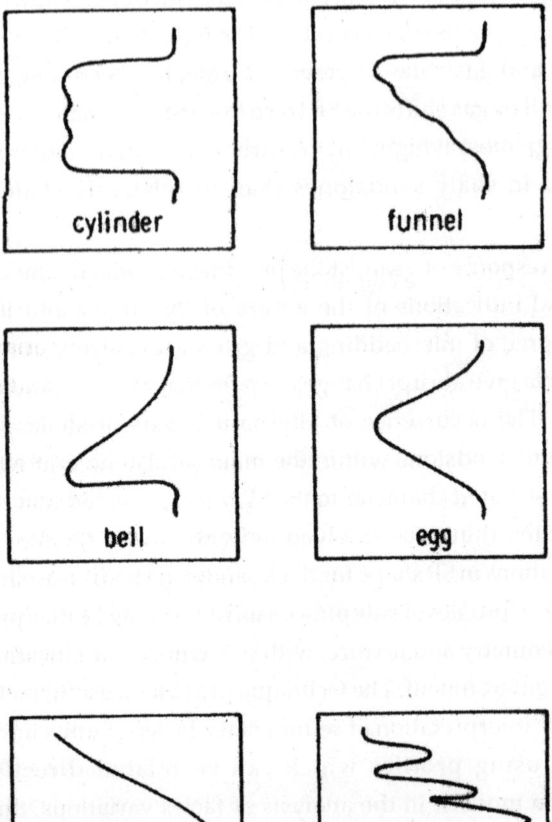

Fig.17. Basic descriptive shapes of SP profiles.

References

1. Asquith, D.O., 1970; Depositional topography and major marine environments, Late Cretaceous, Wyoming; *Am. Assoc. Petrol., Geolog. Bull*; Vol. 54, No. 7, pp 1184-1224.

2. Campbell, R. L., Jr., 1968; Stratigraphic application of dipmeter data in mid-continent; *Am. Assoc., Petrol, Geolog. Bull,* Vol. 52, No. 9, pp. 1700 -1719.
3. Chombart, L.G., 1960; Well logs in carbonate reservoirs; Geophysics, Vol. 25, No. 4, pp. 2779-853.
4. Choquette, P.W., and Pray, L.C., 1970; Geological nomenclature and classification of porosity in sedimentary carbonates; *Am. Assoc. Petrol. Geolog, Bull.,* Vol. 54, No. 2, pp. 207-250.
5. Doveton, J.H., and Bornemann, E., 1981; Log normalization by trend surface analysis; *The Log Analyst,* Vol. XXII, No. 4, pp. 3-8.
6. Doveton, J.H, 1986; *Log Analysis of Subsurface Geology;* John Wiley & Sons, New York.
7. Eaton, B.A., 1969; Fracture gradient prediction and its application in oilfield operations; *J Petrol. Tech.,* Vol. 21, pp. 1353-60.
8. Fertl, W.H., 1979; Gamma ray spectral data assists in complex formation evaluation; *The Log Analyst,* Vol. XX , No. 5, pp. 421-423.
9. Fork, R.L., 1951; Stages of textural maturity in sedimentary rocks, *J. Sed. Pet,* Vol. 21, pp. 127-130.
10. Gilchrist, W.A., Jr., Quirein, J.A., Boutemy, Y.L., and Tabanou, J.R., 1982; Application of gamma ray spectroscopy to formation avaluation; Soc. Prof.Well Log Analysts 23rd Ann-Logging Symp., paper B, pp. 28-30.
11. Gilreath, J.A., and Maricelli, J.J., 1964; Detailed stratigraphic Control through dip computations; *Am. Petrol. Geolog. Bull.,* Vol. 48, No.12, pp. 1902-1910.
12. Goetz, J.F., Prins, W.J., and Logar, J.F., 1977; Reservoir delineation by wireline techniques; *The Log Analyst,* Vol. XVIII, No.5, pp. 12-40.
13. Griffiths, J.C., 1952; Grain-size distribution and reservoir rock characteristics; *Am. Assoc. Petrol. Geo. Bull.,* Vol. 36, No. 2, pp. 205-229.
14. Hill, H.J., and Milburn, J.D., 1956; Effect of clay and water salinity on electrical behaviour of reservoir rocks; *Trans. Amer. Inst. Mech. Eng.,* Vol. 207, pp. 65-72.
15. Jeffries, F.,1965; Wireline logs – the infant oil finding technique; *Can. Petrol. Geol. Bull.,* Vol. 13, No. 2, pp. 291-302.

16. Powers, M.C., 1967; Fluid release mechanisms in compacting marine mudrocks and their importance in oil exploration; *Am. Assoc. Petrol. Geolog. Bull.*, Vol.51, No.7, pp. 1240- 54.

17. Schlumberger, 1972; Log Interpretation : Vol. I. Principles; Schhmberger Ltd., New York, pp. 113.

18. Selley, R.C., 1976; The habitat of North Sea Oil; *Proc. Geolog. Assoc.*, Vol . 87, Part 4, pp. 359-387.

19. Serra, O., and Abbott, H.T., 1980; The contribution of logging data to sedimentology and stratigraphy; Society of Petroleum Engineers Preprint 9270, pp.19.

20. Serra, O., Baldwin, J., and Quirein, J.,1980; Theory, interpretation and practical applications of natural gamma ray spectroscopy; Trans. Soc. Prof. Well Log Analysts Am. Logging Symp., Paper Q, pp. 30.

21. Skelton, J.W., and Mack, D.E., 1970; Gain orientation in determination of paleocurrents and sandstone trends; *Am. Assoc. Petrol Greolog. Bull.*, Vol. 54, No. 7, pp. 1108-1119.

22. Visher, G.S., 1961; How to distinguish barrier bar and channel sands; *World Oil*, May issue, pp. 106-113.

23. Wahl, J.S., Tittman, J., Johnstone, C.W., and Alger, R. P., 1964. The dual spacing formation density log; *J.Petrol. Tech.*, Vol. 16, No. 12, pp. 1411-16.

24. Walsh, J.L, 1923; A closed set of normal orthogonal functions; *American J. Math*, Vol. 45, pp. 5-24.

25. Wilson, M.D., and Pittman, E.D., 1977; Authigenic clays in sandstones; recognition and influence on reservoir properties and paleoenvironmental analysis; *J. Sed. Pet.*, Vol. 47, No. l, pp. 3-31.

Appendix-A

Global Cooperation Within the Petroleum Industry

Introduction

The petroleum companies have gradually moved from national to international companies, first characterised by global operation, subsequently characterized by global cooperation as required by international agreements and national aspirations. International oil companies and service companies had to cooperate in completely new way with host country industries. This has since spread worldwide and been successful in areas of transparent democracies as host nations.

Need for Globalisation—Global cooperation has been forced as a consequence of public pressure to rethink the sharing of benefits and threats from hydrocarbons between companies, national and regional interests. Both environmental and societal concerns have been put in focus, and have become a basis for cooperation globally. It is in the enlightened self-interest of energy companies of today to consider the significance of sharing in a broader sense than what used to be common place thinking only a few years back. The petroleum industry will take on these new challenges while providing affordable and steadily cleaner operations and fuels in increasing amounts for many years ahead.

Three Powerful Global Forces are—globalisation, liberalisation and technology. Unless these three global forces are given a human face, especially in growing economies we shall have an increasing problem to solve. The question of global cooperation being a means to safeguard the interests of host nationals, host nation companies and visiting companies.

It is also appropriate to consider possible future developments on the basis of scenarios. Scenarios offer an opportunity to see different possible developments and consequently new risks and opportunities that a business may face. It is relevant to understand that energy is of fundamental importance for the development of societies and people. The need for affordable and clean energy will continue to increase, especially in growing economics. Both consumption and production of energy is central in all countries, but critical for many. Moreover, it is reasonable to assume that sustainable development also will require cleaner and more effective energy alternatives. Global cooperation is absolutely essential to manage the transition to a substainable energy world. The petroleum industry is crucial in this setting, but so is also the work of international organisations and host governments.

Most energy scenarios project that consumption of every primary energy source will increase over next 20 year forecast horizon. Most of the increament in energy consumption is in the form of fossil fuels (oil, natural gas, and coal), because it is expected that fossil fuel prices will remain relatively low, and that the cost of generation energy from non-fossil fuels will not be competitive. It is possible, however, that as environmental programmes or government policies, particularly those designed to limit or reduce green house gas emissions, are implemented, the outlook might change, and nonfossil fuels (renewable energy sources such as hydroelectricity, geothermal, biomass, solar, and wind power) might become more attractive. But for the future, fossil fuels will dominate (90 per cent) albeit using clean technologies. Overall energy use is closely linked to population growth. Over half of the projected growth in energy demand will occur in tandem with strong economic growth in the nations of Asia including China, India, and South Korea, and in Central and South America.

Start of Globalisation—Today, the oil industry remains as competitive as ever but more readily recognises the value of partnerships in some parts of the business. Innovative joint venture agreements are constantly being developed, both between competing oil companies and with specialist organisations outside the industry.

Joint ventures and collaborative agreements are useful way of minimising risks on a number of fronts. Oil companies have to deal with geological risks associated with prospects, political risks in the countries in which they operate and technological risks in a various parts of their business. By collaborating with others they potentially lose some of the upside but can pool resources and ideas and spread the risks more thinly.

Requirement of Globalisation

(a) **Joint Ventures** : Long established cooperation between oil companies in the upstream business, refining and marketing is being supplemented by new joint ventures and partnerships, both in traditional and developing areas of the industry. For example in the mature oil province of U.K. North sea, large oil companies are using collaborative approaches to make their assets work harder while smaller companies entering for the first time are trying to find economic ways of exploiting ever smaller pockets of oil around existing infrastructure. Talisman Energy, the Canadian exploration and production group, entered into a risksharing contract with Global Marine Integrated Services that enabled it to access some of the most marginal offshore oil fields in the world. Using advanced drilling Global Marine was able to tap into the tiny Beauly field, which has reserves of only 3 million barrels of oil. A horizontal extension of 1200 metres was drilled from an existing waell and this was then tied back through a subsea pipeline to a floating production.

At the other end of the traditional business, oil companies are seeking ways to minimise risks associated with the construction of their retail outlets. The global nature of the large integrated oil groups means they have retail operations in many countries, each with their own complex planning laws and different supplies. BP has sought to simplify this process by outsourcing its $ 3 billion service station construction and maintenance programme in Europe, Japan, Venezuela and the US to Bovis Land Lease, the Anglo-Australian construction group. To enable the risks to be shared, Bovis is contracted on a long-term partnership agreement and derives its profit purely from the saving it achieves. New areas while joint ventures and partnership agreements have helped greatly to streamline traditional activities, the greatest concentration of collaborative working in recent years has been in the non-traditional and developing areas of the business. These fall broadly into two categories, new energy technologies and e-commerce. Major oil companies have realized that they must get involved with the development of renewable and alternative energy forms at an early stage, reasoning that commercially recoverable hydrocarbon reserves will eventually be depleted, and that precautions need to be taken against global warming. Some have had in-house research and capabilities working in this area for years, but recognising that alternative energy is outside their core areas of expertise, are now forming joint ventures with more specialised companies. The development of hydrogen fuel cells, which could one day replace the internal combustion engine, has been one of the most fertile grounds for this. All of the major oil companies have linked up with automotive and fuel cell manufacturers in a bid to share risk and ideas.

(b) **Social Concerns**—By demonstrating its ethics and integrity, the oil and natural gas industry safeguards its reputation and credibility. Through this, a respect is shown for people, communities, and the environment, safety and compliance with the law. It is the right thing to do. The global demand for energy and more specifically clean energy is growing rapidly. There is no doubt that energy is essential to the improvement of human life and economic development. Oil, gas and coal represent around 90 per cent of commercial energy used worldwide and technological innovation and good stewardship will allow us to go on finding and producing oil and natural gas for many years past current predictions.

To meet this demand, the petroleum industry is called upon not just to provide effective management of oil and natural gas reserves, but also to marshal its financial resources and technical capacity to further diversify energy sources and delivery options. Energy industries have always evolved to meet people's changing needs. The energy industry of the 21st century will become more diverse and more dynamic. For the longer term, sustainable economic growth, meeting global energy demands in an environmentally sensible and socially acceptable way will require the development of entirely new energy technologies and other advanced fuel sources.

Today, business success, in both developed and developing countries, is increasingly linked to strong performance in financial markets, environmental stewardship and community affairs. Businesses also have to deal with changing sets of expectations from a broader range of stakeholders. The international petroleum industry takes this challenge very seriously. The actual responses to these environmental and social challenges differ among companies and countries. Specifically, the international petroleum industry is collectively striving to address society's need by helping to develop solutions and responses that are socially, economically and environmental acceptable by: (l) Developing partnership with key UN bodies, global organisations and other industry associations (2) searching for cost effective solutions based on sound science, and (3) conducting consultations with all stakeholders.

There is great similarity among the policy level environmental commitments and general aspirations of most major oil companies. These common elements provide an umbrella framework with the needed flexibility for each company to implement it according to its core values. Major oil and gas companies are committed to the sharing of good practices and learning from each other so that they can continuously improve their environmental performance.

(c) **International Cooperation**—Many petroleum led Air Quality Management Programmes within Europe and Scandinavia are actively promoting policies and practices that are objective, cost, effective, and based on sound science, to improve air quality in rapidly developing cities around the world. No matter how good preventative measure are, the petroleum industry recognizes that accidents can still happen. Particular emphasis has been placed on the developing countries where resources and expertise are scarce. Some initiatives coordinate on the ground activity to help the government and the oil industry in specific countries to join forces and develop a credible response and planning strategy. The resources in the oil industry and the specialist expertise available from both the industry and the institutes are used to maximum effect in facilitating national and regional workshops and training programmes attended by both government and industry personnel.

Climate change is a global environmental concern with potentially significant consequences for society and the whole fossil fuel industry. Its significance stems both from possible future impacts of changing world climate as well as potential socio-economic consequences of policies proposed to respond to increased concentrations of greenhouse gases in the atmosphere. A variety of technical options are available which could reduce emissions, the most common and easiest to implement quickly being improved energy efficiency. Other options include fuel switching (coal or oil to gas) renewable energy sources, alternative energy economies (methanol, hydrogen) and nuclear power. Alternatively, greenhouse gases especially CO_2, may be recovered and stored. Utilisation of CO_2 to make chemicals could use only a small proportion of the amount that could be captured. The first CO_2 saline reservoir storage project has now been in operation for over five years, commissioned offshore in the second half of 1996 by Statoil, the operator of the Sleipner field in the North Sea off the coast of Norway. Statoil on behalf of a group of partners operates the Sleipner field. A collaboration of 30 countries has been working on practical research in this area since 1991.

(d) **Transparency**—The petroleum industry has been operating in some of the world's most sensitive environments for more than 100 years. It has been a pioneer in developing and using new technologies, systems and practices to minimise the impact of its operations. With careful planning, management and consultation, energy resources can be developed without causing lasting damage to the environment, while maximising economic and social benefits to the local communities. In order to be successful, oil companies need access to resources and a licence to operate. Businesses have to deal with changing sets of expectations from a broader range of stakeholders. Establishing and maintaining good relationship with all stakeholders in key to the future of the oil industry. It is essential to effectively engage our community stakeholders in decision-making processes that affect them. A trusted company can be part of the process and thus influence the outcome. As trust diminishes, the demand for transparency and assurance increases. Stakeholders went demonstrable assurance that companies are behaving responsibly. They want to be consulted on issues of concern to them. Financial performance and responsibility to shareholders as well as acceptable social and environmental performance is of equal key importance to sustained business success.

Companies such as Statoil, Shell and BP have broken new ground not only in assessing the social impact of new projects but in continually measuring their impact over the life of such projects. Transparency has also been greatly improved in recent years with the introduction of annual reports, which provide unprecedented detail into how companies operate. Major oil companies have realised that they must get involved with the development of renewable and alternatives energy forms at an early stage, reasoning that commercially recoverable hydrocarbon reserves will eventually be depleted, and that precautions need to be taken against global warming. Some have had in-house research and capabilities working in this area for years, but recognizing that alternative energy is outside their core areas of expertise, are now forming joint ventures with more specialised companies.

Conclusion—Meeting society's increasing demand for energy and strong environmental and social performance in this century will be a decisive challenge for energy companies and host nations. The petroleum industry recognizes the complexities of trying to achieve a balance among economic, environmental and social expectations. There are conflicting demands and contradictory expectations. There are limits to what companies can do, and ought to do. At the international level, organizations such as the World Petroleum Congress encourages dialogue and sharing of experiences between its members internal discussion and production of case studies, which all help in building on key lessons learned and sharing good practices within the industry, thereby living up to its purpose to promote sound fossil fuel use for the benefit of mankind.

Appendix-B

Energy Requirement Scenario of India

Oil account for over a third of India's total energy consumption, coal over half and gas, hydro and nuclear the remainder. Future oil consumption in India is expected to grow rapidly, from 3.0 to 3.4 million barrel per day by 2010. India now permits foreign involvement in exploration, an activity long restricted to Indian state owned firms. The largest winner in the bidding round was India's domestic Reliance Industries, in partnership with independent Niko Resources of Canada, which received 12 blocks. British independent Cairn Energy, Russia's Gazprom, the U.S. firm Mosbacher Energy, and Geopetrol of France were all awarded single blocks in partnership with Indian firms. India's state owned Oil and Natural Gas Corporation (ONGC) was awarded eight blocks, three of which it will hold in partnership with other public sector Indian firms. A second round of bidding, with a total of 25 blocks offered, concluded in March, 2001. Sixteen of the blocks have been awarded to ONGC, and four blocks to Hardy Oil of the United Kingdom, in partnership with India's Reliance Petroleum.

For most of the 1990s, India imported a large quantity of refined products. In 1999, refinery construction allowed India to close the gap. At the end of 2001, India had a total of 2.1 million barrel per day in refining capacity, an increase of 9,70,000 barrel per day since 1998. In 1999, Reliance Petroleum's huge Jamnagar refinery came on stream. It has since reached its full capacity of 5,40,000 barrel per day. Another major downstream infrastructure development is the construction of pipelines being undertaken by Petronet India, a company created by an agreement in 1998 between India's state-owned refineries, which will add 5,00,000 barrel per day to India's current 3,25,000 barrel, per day capacity for pipeline transportation of refined products. Pipelines between refineries and major urban centres will replace rail as the main mode of transportation.

Several multinationals have entered the Indian lubricants market, which was deregulated. Over one-third of the market is currently held by such firms as Shell, Exxon Mobil, and Caltex. While these operations are relatively small, they are seen as allowing the majors to study the Indian market, establish brand recognition, and prepare for the eventual deregulation of the Indian retail petroleum products sector.

In a restructuring of the state-owned oil sector, two of the main firms, Indian Oil Corporation (IOC) and Oil and Natural Gas Corporation (ONGC) formed a strategic alliance in early 1999 and swapped 10 per cent their respective shares. As ONGC is an upstream producer and IOC is a downstream refining and distribution firm, the stated aim of the alliance is to create an entity which can compete with the major multinational oil firms.

Indian consumption of natural gas has risen faster than any other fuel in recent years. From only 0.6 trillion cubic feet (Tcf) per year in 1995, natural gas use was nearly 0.8 Tcf in 1999 and is projected to reach 1.3 Tcf in 2005 and 1.8 Tcf in 2010. Increased use of natural gas in power generation is to account for most of the increase, as the Indian government has been encouraging the construction of gas-fired electric power plants in coastal areas where they can be easily supplied with liquefied natural gas (LNG) by sea.

India is investing heavily in the infrastructure required to support increased use of natural gas. Gas Authority of India Limited (GAIL), a government-owned entity, is to undertake a doubling of capacity on its main Hazira, Bijaipur, Jagdishpur (HBJ) Pipeline. GAIL also plans a new distribution network in West Bengal and a pipeline which would connect Kolkata with Chennai. Shell has signed a memorandum of understanding (MOU) with the state government of Uttar Pradesh in northern India for the development of gas distribution infrastructure.

Shell has received approval for an LNG import terminal at Hazira in Gujarat, and has contracted for LNG supplies from Oman. Reliance industries also plans an LNG import terminal at Jamnagar in Gujarat, neat its oil refinery. Aside from LNG imports, imports of natural gas by pipeline may play a role eventually in satisfying India's gas needs. One possibility would supply India with natural gas from Iran's huge South Pars field via pipeline. Australia's Broken Hill Proprietary (BHP) is the main forein backer of the project. An offshore route bypassing Pakistan is under study by Snamprogetti of Italy .

Coal is currently the dominant commercial fuel in India, satisfying more than half of India's energy demand. Power generation accounts for about 75 percent of India's coal consumption, followed by heavy industry. Coal consumption is projected to increase to 427 million tonnes in 2010, up from 348 million tonnes in 1999. India is the world's third largest coal producer, after the China and the United States, so most of the country's coal demand is satisfied by domestic supplies. Increased coal consumption over the past four decades has led to a nine-fold increase in energy-related carbon emissions. Environmental effects due to the relatively high use of coal in the energy mix are exacerbated by the low energy efficiency of coal based electricity generating plants. Inefficient plants are one of the contributing factors to a steadily increasing energy consumption per unit of output. With the high costs associated with replacing existing coal based plants, it is realistic to assume that these plants will continue running to the next couple of decades.

India's per capita energy use and carbon emissions, while lower than the world average, result in a substantial percentage of world energy use and carbon emissions, due to the country's large population and heavy reliance on coal. Increased use of renewable energy is one means of reducing carbon emissions. Two major sources of renewable energy in India are wind power and hydroelectric plants. India's five-year plan for 2002 -2007 calls for 10 per cent of new electric generating capacity to come from renewable sources.

Future energy needs are shaped by four things : (1) demography (2) urbanisation, (3) incomes, and (4) liberalisation. India has the second largest population today, about one billion. By 2050 this could grow to 2 billion. Imagine the challenges when considering urbanisation, incomes and liberalisation. Total energy consumption will, double or triple in fifty years, and the global population could grow from 6 billion at present to 8.5 billion. India will depend very much on extremely energy efficient energy markets and alternatives. It seems that renewable energy and hydrogen could be likely contenders. Natural gas could then be a bridge fuel in next two decades. A number of scenarios suggest that we will evolve our energy supplies moving from high to low carbon fuels and towards electricity and hydrogen as the dominant energy barriers. However, as reserves and resources become an issue bio-fuels are expected to grab a larger share of the energy market.

Appendix -C

National Oil Companies

1. Oil and Natural Gas Corporation (ONGC) limited

Introduction :

ONGC is one of the few oil companies in the world engaged in every aspect of oil exploration, drilling, production, research and development. It has the rare experience of exploring for oil in the most diverse geographical conditions. It operates both on shore as well as offshore and, also has overseas operation. It has developed a high degree of expertise and carries out most of the operations itself. Till date, ONGC is Indian's leading "Navratna". ONGC is exploring new frontiers worldwide. Today, ONGC is the world's sixth largest integrated Exploration and Production Company with a market capitalization of over Rs.17039.50 crores and annual revenue generation to the tune of Rs. 15101 crores during 2000-2001.

Since the demand for the oil is ever increasing and as the proven onshore reserves of India are gradually dwindling, it had become necessary for ONGC in 1975 to explore for oil in the offshore areas apart from the intensification of onshore exploration. The potentially rich offshore area extend for 5700 kms. along the entire length of Indian coast. The total area of the continental shelf upto a depth of 200 metres of water is approximately 400,000 sq. kms.

Now ONGC is planning to do the marine geophysical survey in the deeper water. Currently, India follows the 400 m bathymetry as the initiation of deepwater. Deep water exploration has been guided by the inventions of technology to build a host of submersible drill ships with increased abilities to be stable and withstand the harshest invironmental conditions. ONGC's management hedged it risk planned upgradation of an existing drill ship, Sagar Vijay, to enable it to drill in water depths upto 900 m. This was successfully accomplished in 1998. ONGC is also upgrading another drilling ship, Sagar Bhushan, to enable it to drill upto a water-depths of 400 m.

From 1956 to 2001, ONGC has found 6 billion tonnes of oil and natural gas (Initial in place oil & oil equivalent) in India. In the next 20 years (from 2001 to 2020), ONGC is aiming to find another 6 billion tonnes.

Objectives of Formation

The world's first oil well was completed in Pennsylvania in 1859. The Burmah Oil Company showed India the way and put India on the oil map of the world in 1866. India's first oil well was successfully drilled at Nahorpung in Upper Assam and had been struck in 1890. At that time technical know how for the production of oil was not available within the country.

In 1955, an Indian delegation visited some European countries including the Russia to make a study of the oil scene there. Government of India formulated a national policy according to which development of oil resources was made the exclusive responsibility of the state.

As a first step towards the implementation of this policy, a petroleum division was created within the Geological Survey of India in October, 1955 exclusively for oil exploration in the country. In December, 1955 Soviet Oil experts came to India and submitted a report to the Government about oil prospecting.

Soon, the petroleum division of Geological Survey of India grew into the Directorate of Oil and Natural Gas which was raised to the status of commission on 14 August, 1956. The Commission was later converted into a Statutory Body on 15 October, 1959, with its headquarter at Tel Bhavan, Dehradun.

There were two stalwarts who were perennial sources of inspiration for the Oil and Natural Gas Commission. One was Pandit Jawaharlal Nehru who gave support and encouragement to the geoscientists of this organisation. Another was Shri K.D. Malaviya who was the minister in the Nehru's Cabinet. He was the real architect of India's petrochemical policy and founder of petroleum industry.

Oil and Natural Gas Commission came into being at a time when most of the Oil wealth of India was undiscovered. At that time entire indigenous production of oil was from Upper Assam area and it was just 3,96,000 tonnes.

The ONGC (Oil and Natural Gas Commission) on 1st February, 1994 became a public limited company vested with greater financial autonomy to compete better in keeping with liberalisation of the economy. Its Chairman is redesignated as Chairman-cum-Managing Director.

The 34-year old Oil and Natural Gas Commission Act stood replaced and all its assets, liabilities and personnel were transferred to the new Corporation. By becoming a public limited company under the Companies Act (1956), the Oil and Natural Gas Corporation has been allowed to go to the Capital market to dilute its equity to the extent of percent by offering expanded equity to the public mutual funds and the employees of the Corporation.

During its growth and development, ONGC had the benefit of having illustration of Chief Executive like Mr . S. S. Khere, Mr. P. R.. Naik, Mr. A. Zaman, Mr. L.J. Jonson, Mr. B.S. Negi, Mr. N. B. Prashad, Mr. P.T. Venugopal, Col. S.P.Wahi, Mr. S. L. Khosla , Mr. S.K. Manglik and Mr. B.C. Bora. Mr. Subir Raha took over as the Chairman and Managing Director of Oil and Natural Gas Corporation with affect from 25 May, 2001. These Chief Executives of ONGC have shaped the destiny of the organisation through their committed efforts. Today, almost in every area of operation, ONGC has performed with remarkable efficiency.

Oil Needs Integrated Approach

Oil is the life line for the industry and no industrial progress can be possible without it. For India, oil and natural gas play a pivotal role in the speedy progress. The Oil and Natural Gas Corporation and Oil India Limited have done commendable jobs in not only finding new oil fields but also in successfully developing them. Their efforts so far give an optimistic picture about indigenous oil prospects. India's aim is to increase oil and natural gas production quickly and thus save valuable and limited foreign exchange.

The multi-pronged strategy of ONGC envisages integrated exploration, drilling and production activities. It is marshalling its resources to meet some of the fundamental challenges of exploration. One of them is to convert a considerable quantity of geological resources into exploitable reserves so that the hydrocarbon potential of the country is upgraded.

Oil has been traditionally considered an integrated business. Leading internation oil companies are integrated from the exploration stage right upto refining and marketing. In India the first and foremost task is to have oil industry completely integrated. After integration, the industry should be given the directive to make the country self-sufficient and self-reliant in oil and natural gas within a specified period of giving it a free hand.

The search for oil is a continuous process and unless more oil is located as far as output rises, the reserve production ratio will necessarily tend to decline as years go by. There is still a good deal more oil in the earth than has been taken out.

Foreign capital and expertise should be welcomed in our search for oil and natural gas. It is a right decision to invite foreign oil companies to help us in the search.

An aeromagnetic survey of Western Rajasthan, Punjab, Himachal Pradesh and Uttar Pradesh was undertaken by ONGC and began drilling of wildcat wells. In May 1958, gas was struck at Jawalamukhi and in September the same year at Cambay. In 1960, oil was struck at Ankleshwar, followed by a similar discovery in Rudrasagar in Assam in the concession area which Burmah Oil Company had surrendered to the Government of India. Ankleshwar and the North Gujarat Oil fields had proved most productive. In Assam, earlier ONGC discoveries were in former Burmah Oil Company and Assam Oil Company exploration licence area.

It was Soviet geophysical marine survey ships that explored India's West Coast in 1964, for a distance of about 200 km from Bombay Port. That area was later named Bombay High. The total area of Bombay High is about 1,43,000 Sq.Km. India drilled her first offshore well at Aliabet from a fixed platform in shallow waters. Second offshore well was drilled in Tarapore structure by using mercury type jackup offshore drilling rig named "Sagar Samrat". The second well drilled by this rig struck oil in 1974. Since then ONGC, side by developed and produced oil from this structure and the satellites around it. Oil struck in Ratnagiri, Godavari offshore and portonovo in 1980.

In the 1980s Russia had helped India to set up a geophysical service for vertical seismic profiling of wells and had trained Indian Geo-Scientists in the use of shear waves to boost the efficiency of geo-physical works. Joint works had been undertaken in India to assess the parameters of productive formations. The Soviet had developed a downhole motor engine which had revolutionised offshore drilling.

Experts of the two countries were hopeful about the success of the ambitions Indo-Soviet programme for oil development. The well-known Soviet expert, Prof. N.A. Kalenin, had expressed confidence that India would be able to attain self-sufficiency in oil and natural gas by the early 21st century. Prof. Kalenin prepared the first comprehensive report of India's oil and natural gas reserves in the mid-fifties. He had stated that India could give a big boost to her oil production through offshore exploration. According to his opinion India has considerable oil and gas reserves on land and offshore. These oil and gas reserves lie at accessible depths.

ONGC Towards Meeting Future Challenges

Oil business is a high tech activity involving the vagaries of the luck factor and the ever changing conditions of the world market. The phenomenal growth witnessed by the petroleum industry in the last four decades is unprecedental in the Indian industrial context. The oil industry was at a very nascent stage when India achieved independence. The future augurs well for ONGC. The infrastructure and facilities at the disposal of ONGC enable it to perform the way it does and elevate it to the international standards making it one of the top 25 oil majors worldwide. Today, the onshore systems include 225 production installations, 7900 kms of pipeline network, 2 major onshore terminals, 75 drilling rigs, 66 workover rigs, 45 seismic, 36 logging units while the offshore systems include 131 well platforms, 5 well-cum-process platforms, 28 process platforms, 18 drilling and jack-up rigs, 3200 kms of pipeline network, 33 offshore supply and 4 special application vessels.

ONGC has taken conscious decision to undertake ventures in other parts of the globe. Besides, providing it an access to outside would and spreading its risk this also helps ONGC with an exposure to the prevailing level of technology in other parts of world. ONGC has already worked in Vietnam, Tanzania, Iraq, Iran, etc.

By any yardstick, cost of operation, productivity, net reserves and surpluses, gross profits generation of internal resources and ability to meet targets, ONGC can be singled out as a model enterprise for other industrial enterprises to emulate. The oil industry is often referred to as imput deterministic and output probabilistic. Perseverence and the ability to take calculated risk are the two essential ingredients to sustain growth in oil exploration.

Now the oil exploration and production technology have grown increasingly advanced and complex. The drilling of high pressure/high temperature wells in onshore fields, application of secondary and enhanced recovery methods, horizontal well-drilling and subsea completion multiple completion technology and underground coal gasification processes are few instances of the strides in technology operation. ONGC has kept pace with the sweeping winds of change in oil technology and has been adopting successfully the latest state-of-the-art technology. Also it is innovating to adopt to the specific local needs. To-day, India has one of the very few oil companies in the developing world to possess a rare degree of technological self-reliance on par with the latest advanced technology available in the other parts of the world.

By drilling few horizontal wells in Bombay High, ONGC has joined the selected group of international companies possessing this specialised technology. So far, only few dozens horizontal wells have been drilled worldwide.

It is not enough to perform well and plan out strategies to meet future challenges. It is equally important to create awareness both within the organisation and outside for obtaining desired momentum.

2. Oil India Limited (OIL)

Introduction

Oil India Limited is a Government of India Enterprise engaged in the exploration, drilling and production of crude oil and natural gas from defined concessional areas in Upper Assam/Arunachal Pradesh and in eastern northern and western India. Although her entry into the public sector in just 21 years, the roots of her experience stretch back through antecedent companies to the beginning of this century. In the blood stream of OIL runs the courage and daring and professional integrity of the early pioneers. These qualities are shared in full measure by the Oil and Natural Gas Corporation (ONGC) limited. Both Oil and ONGC are conscious of the tremendous responsibility thrust upon them by an oil-starved nation. If their past is any index, they can be trusted to give of their best in fulfilling this responsibility in the coming years.

The story starts at the beginning of the 19th century, with a steamy jungles of Upper Assam. It recounts the valiant efforts of individuals and organizations to find oil under conditions of extreme physical hardship. These efforts climaxed in the discovery of the Digboi Field which passed into the hands of the Burmah Oil Company Ltd. (BOC) in 1921 and to Oil India Limited (OIL) in 1981. Despite very low production, Digboi Oilfield retains the distinction of being the world's oldest continuously producing oilfield.

Search for Oil : The rapid expansion of geophysics during the past few decades has opened up a new era of state-of-the-art technology for hydrocarbon exploration and development. However, the story of oil exploration in India began in dense jungles and swamps in the extreme north-eastern corner of Assam in the early years of 19th century. The petroleum industry in India started off in 1880 with the discovery of Digboi oil field in Upper Assam. A tower was completed at Digboi. It was a modest, thatch-covered wooden structure about 20 metres high with little geometry and less aesthetics. It was the tower, or derrick, of Digboi Well No. 1.

Digboi Oil was discovered by the antecedent companies of Oil India Limited, from the direct evidences of seepages and out crops. The evidence of oil in foothill areas led to the speculation that oil might be also present in the adjacent areas at much deeper levels. To study hydrocarbon occurrence in these deeper horizons, some new technique was felt necessary such as geophysical methods, which were deployed in those days to establish many medium to large size oil and gas fields.

The geophysicists in Upper Assam forged ahead from torsion balance surveys in 1925 to the area of gravity, magnetic and seismic surveys and efforts have been made to enhance the exploratory activities utilising these methods in order to establish hydrocarbon potential. Government of India has taken many steps to accelerate the pace of exploration and production in the country. Geophysics has been playing an important role right from 1960s with the invention of digital field systems for data acquisition and computer for processing. OIL has been adopting such geophysical technique from time to time.

In 1937 Geophysical License was issued by Assam Government for giving the exclusive rights to carryout geophysical surveys for a period of 5 years. In accordance with the terms of the geophysical licence, the Assam Oil Company begain gravity and seismic survey of the alluvial area of Upper Assam. Seismic surveys were undertaken by Assam Oil Company Ltd. in 1938 and suggested a Seismic High at Nahorkatiya (Assam). It was only in May 1952 that drilling could be taken up on this structure and this resulted in the completion of Nahorkatiya-1 well as an oil producer. Immediately after discovery of NKH-1 oil well, gravity survey of the whole alluvial area using the improved methods for aeromagnetic survey and seismic surveys in selected areas was carried out. The seismic surveys were carried by a British firm called Seismograph Services Ltd. (SSL). A number of structures were identified for drilling. A structure was delineated which resulted in the discovery of Moran oilfield in October 1956.

After incorporation of Oil India Limited in 1959, the work of data compilation and interpretation was continued and seismic data acquisition work was resumed in 1963 by hiring the services of SSL during 1963-64 and ONGC in 1964-65. During these years, the surveys were also extended to the concessional areas falling in state of Arunachal Pradesh. Seismic surveys were carried out every year from 1969 to 1974 by hiring seismic party from ONGC. During the late past of seventies, OIL drew up a plan to undertake in-house exploration activities in Assam & Arunachal Pradesh. OIL's furture thrust in the northeast for geophysical exploration will be on frontier areas. Areas identified for extensive exploration in the near future are logically difficult, e.g., thrust belt, areas covered with boulder beds and thick forests, and those lying under the Brahmaputra river bed.

Realising the importance of geophysics for search of oil and gas, OIL has starts taking concessional area outside Assam for extending its exploratory efforts outside Assam. A project was established as Bay Exploration Project at Bhubaneswar (Orissa) during the year 1977-78. For the first time, during 1978, OIL had carried out Aeromagnetic surveys through NGRI and seismic offshore surveys by Delta Exploration Inc. of USA in Mahanadi Offshore areas. Exploration for hydrocarbons in Rajasthan was initiated by ONGC and later joined by OIL for carrying extensive geophysical work in the Jaisalmer and Bikaner-Nagaur Basin when PEL over an area of 28600 sq.kms. was granted in 1983. OIL started its exploration activities in 1984. Geophysical exploration in Saurashtra offshore initiated during late eighties with the grant of 26000 sq.kms. areas in offshore. The Saurashtra Exploration Project was established in 1989 at Rajkot (Gujarat) and the seismic survey in offshore areas commenced subsequently. OIL has diversified its activities for exploration in the area of Ganga Basin after obtaining the PEL for 13,000 sq.kms. area known as Kashipur PEL area. The seismic data acquisition commenced in December 1990. The seismic surveys in the Ganga Basin by OIL are still in progress in order to identify the future drillable prospects in the area.

OIL's major produetion today is from Eocene reservoirs. Future activities, therefore, will also focus on the exploration of these Eocene sands. There reservoirs are thin, subtle and deeper than the productive Barail sands of the Oligocene age. These deep and subtle sand bodies will also demand a radically different approach to exploration. Further, as the shallower Eocene prospects reach a mature state in their life cycle, the focus of exploration within the Eocene will progressively shift towards deeper horizon. OIL intends to leverage its 3D seismic capabilities to discover and develop these subtle reservoirs. In future, OIL has a plan to cover all prospective areas in South bank by high quality 3-D seismic survey.

Achievements

Petroleum Geophysics in OIL has played a vital role in identifying the drillable prospects not only in the areas of Northeast but also in western part of the country including Mahanadi onshore offshore areas for exploration.

A total of about 11,000 sq. km. of area has been covered by Aeromagnetic surveys in Assam and Arunachal Pradesh and 13,000 sq. km in western U.P. In total, a total of 56 per cent areas of Assam and Arunachal Pradesh has been covered by seismic alone and 44 per cent of the area is yet to be covered out of the total PEL area held by OIL. About, 96,000 SLKM of 2-D seismic data and 1100 sq. km. of 3-D seismic data have been processed, reprocessed in-house till date.

Appendix- D

Chronological Development of National Petroleum Industry

1825 Earliest recorded reference to oil in Upper Assam in the despatches of Lieut R. Wilcox of the 46th Regiment Native Infantry.

1825-1845 The era of individual intrepid explorer; C.A. Bruce, Major A. White, Lieut Bigge, Capt. F. Jenkins, Capt. P.S. Hannay: who stumbled upon oil in the course of their reconnaissance surveys.

1865 H.B. Medlicott of the Geological Survey of India while visiting the coal areas of Assam observed the oil seepages at Makum Pani and recommended test drilling in the Makum-Namdang area.

1866-1869 The above drilling plus drilling of test wells in the Jaipur area were carried out by Mr Goodenough of Mckillop, Stewart and Co., Calcutta. Drilling at Makum -Namdang was distinctly more encouraging than at Jaipur, but owing to transport difficulties the venture folded up. One of the above wells at Makum drilled to a depth of 118ft on 26 March 1867, was Asia's first successful mechanically drilled oil well.

1881 The Assam Railways and Trading Co Ltd (AR & T Co.) was registered on 30 July with objectives that included construction of railways and exploiting the natural resource of coal, timber, tea and petroleum.

1889	Spudding in of Digboi Well No. 1 in September by the AR & T Co.
1890	Completion of Digboi Well No. 1 as producer in November.
1891-92	Continued drilling in Digboi field with fair success.
1893	A small refinery erected at Margherita where, for about six years, oil from Digboi was sent by rail: The Assam Oil Syndicate was granted a mining lease over the northern Digboi area.
1899	The AR & T Co. forms a new company, the Assam Oil Company, to take over their oil interests.
1900	Construction of Digboi Refinery begins.
1901	Digboi Refinery commitssioned with running of the first 'still' in December.
1911-12	The Burmah Oil Company Ltd (BOC) arrives on the Indian scene in the Surma Valley.
1915	BOC acquires the oil interests of the Budderporc Oil Company and starts exploration/development of the Badarpur structure.
1919	AOC began drilling at Namchik and abandoned the well the following year at 592 ft.
1920	BOC drilled Masimpur Well 2 and Hilara Well No.1 in Surma Valley. Both abandoned on account of feeble shows.
	AOC drilled a well upto 715 ft. at Balijan, Upper Assam, with no success.
1921	By arrangement with the AR & T Co. the Burmah Oil Company took over in a phased manner the control of the Assam Oil Company operations.
1922	BOC encountered a succession of failures at Chhatachura, Masimpur and Kanchanpur in the Surma Valley.
1923	Wells drilled at Barjan and Namchik in Upper Assam Valley proved to be failures.

First well at Patharia in the Surma Valley was spudded in by the BOC in 1923 and finished in 1931. Recovered about 10,000 barrels of oil before abandoning well.

Digboi refinery entirely rebuilt and capacity enlarged.

1925 Torsion balance survey in Bordubi area represented India's first attempt to use geophysics in the search for oil.

Drilling at Baragolai gave encouraging results, but the well at Dhekiajuli was a failure.

BOC's Masimpur Well No. 4 proved to be a dry hole.

1926 AOC drilled two test wells at Dilli, Upper Assam, to depths of 439 ft. and 328 ft. Unremunerative.

Two product pipelines were laid from Digboi to Tinsukia.

1928 Wax extraction plant installed in Digboi Refinery.

1930 BOC drilled a well to 2757 ft. at Bandersulia and abandoned it the.following year owing to disappointing results.

1931 BOC's Well No. 5 at Masimpur encountered a high pressure gas sand at 5800 ft. capable of yielding 3000,000 cubic ft. a day.

Edeleneau Plant, and Dubbs cracking unit installed in Digboi Refinery.

1933 Badarpur Field in Surma Valley abandoned.

1936 Masimpur Well No. 5 abandoned at a depth of 7685 ft.

1937-38 Extensive programme of geophysical surveys implemented in Upper Assam.

Two cable-tool wells completed by AOC at Makum to depths of 685 and 837 ft. without success.

1939-1944 War time regulations permitted only extension drilling to prove eastern limits of Digboi Field.

1945 AOC drills 2 wells at Balijan to depths of 2428 and 1324 ft.

1948 Geological Survey of India starts geophysical surveys in the Cambay area and delineates the basin.

1949 AOC abandons its efforts to develop the Makum-Namdang Field.

A geological and geophysical survey of the alluvial area of West Bengal and East Pakistan begun by the Standard Vacuum Oil Company (SVOC).

1951 On basis of earlier geophysical surveys, site for drilling selected on the south bank of the Burhi Dihing River at Nahorkatiya.

AOC starts test drilling at Nichuguard in the Naga Hills.

1952 Well No. 1 at Nahorkatiya spudded in by AOC on 26 May.

1953 Gasoline Extraction Plant commissioned in Digboi Refinery.

Nahorkatiya Well No. 1 completed as producer at a depth of 11715 ft. in June.

1954 Lube oil distillation unit, first of its kind in India, installed in Digboi Refinery.

Favourable subsurface structures discovered by AOC at Rudrasagar and Geleki.

Second refinery in India owned by the Standard Vacuum Oil Company comes on stream at Trombay near Bombay.

Indo Stanvac Project for oil exploration in West Bengal formed between Government of India (GOI) and SVOC.

1955 The Oil and Natural Gas Division set up by the GOI.

Third refinery owned by Burmah-Shell Refineries Limited set up at Bombay.

1956 Discovery of Moran Field by the AOC. The first well drilled up to 13,739ft was deepest in Asia; struck oil in November.

Oil and Natural Gas Division converted into a Directorate in May and later to a Commission in August.

1957 Fourth Refinery in India owned by Caltex Oil Refinery (India) Ltd. comes on stream at Visakhapatnam on 15 April.

1958 Promotion Agreement signed between AOC/BOC and GOI for formation of Oil India Private Limited to exploit the oil resources found by AOC in Nahorkatiya and Moran.

Gas struck in Jawalamukhi.

Cambay Well No. 1 (Lunej structure) spudded in by ONGC and completed as a producer.

Indian Refineries Limited, a public sector undertaking registered on 22 August.

1959 Oil India Private Limited incorporated on 18.2.59 and registered at Shillong as a Rupee Company with two-third shares owned by AOC/ BOC and one-third by the Government of India.

ONGC becomes an autonomous statutory body.

1960 Oil struck at Anklesvar in Gujarat and Rudrasagar in Assam by ONGC.

Indian Institute of Petroleum comes into being.

1961 By a new agreement dated 27.7.61 GOI and BOC become equal partners in OIL. Completed 100th well in OIL areas.

India's first gas turbine for power supply commissioned at Duliajan.

Oil struck in Kalol by ONGC.

1962 The first public sector refinery comes on stream at Guwahati.

Nahorkatiya-Guwahati pipeline commissioned on 26.4.62.

First shipment of Anklesvar crude leaves for Bombay by tank wagons.

Oil completes India's first dual completion well on 13.10.62 (NHK Well117).

1963 World's first crude oil conditioning plant commissioned at Nahorkatiya.

Oil drills India's first deviation well (NHK 122).

Offshore seismic surveys started by ONGC in Gulf of Cambay.

Oil struck in Sanand and Nawagam in Gujarat by ONGC.

1964 Guwahati-Barauni crude oil pipeline commissioned.

First long-distance product pipeline in India commissioned from Guwahati Refinery to Siliguri.

Indian Refineries Limited and Indian Oil Company merge to form Indian Oil Corporation Limited.

Oil discovered in Lakwa by ONGC.

Supply of gas from Cambay to Dhruvan Thermal Power Station starts.

1965 OIL pioneers gas injection for pressure maintenance in one of its Nahorkatiya reservoirs.

OIL starts gas supplies to ASEB on a regular basis.

ONGC's 16 ft dia pipeline from Anklesvar to Koyali Refinery commissioned.

1966 For the first time in India AOC applies the technique of formation fracturing in its field to stimulate production.

Cochin Refinery comes on stream.

1967 Water injection starts in Nahorkatiya (OIL) and Anklesvar (ONGC).

1968 Lubrizol India Ltd. starts production.

Oil discovered in Geleki by ONGC.

Oil starts drilling in Arunachal Pradesh at Kharsang, 9th May.

1969 Madras Refinery comes on stream.

ONGC completes drilling of 700 wells.

Discovery of oil in Kusijan by OIL.

1970 After a lapse of 10 years AOC resumes drilling in the Digboi oil field and 12 infiller shallow wells drilled using a light mobile rig.

India's first offshore well in the Gulf of Cambay spudded in on 14 March.

1971 Several new structures discovered in Gujarat by ONGC.

300 wells completed in OIL areas.

1972 AOC starts water injection in Digboi Field.

OIL discovers oil in a new structure at Jorajan.

Highest Wax production in Digboi Refinery: 45852 tonnes.

1973 ONGC completes drilling of 1000th well.

OIL discovers gas in the Eocene at Tengakhat.

1974 First offshore well on the Bombay High spudded in by the ONGC on 31.1.74 using the drill-ship Sagar Samrat; oil struck on Feb. 19. GOI takes over 74 per cent shares in ESSO.

1975 Fuel sector of Haldia Refinery goes into commercial production.

OIL builds first aerial ropeway across the Noa-Dihing in Arunachal Pradesh and initiates India's first polymer flood project.

1976 Bombay High put into commercial production by ONGC

OIL spuds Manabum-1, the first well on the north bank of the Noa Dihing River.

	OIL strikes oil in Kharsang and puts Arunachal Pradesh on oil map of India.
	GCI takes over 100 per cent shares in Burmah Shell Refineries which is renamed as Bharat Refineries Limited and remaining 26 per cent shares in ESSO which is renamed as Hindustan Petroleum Company Limited.
1977	Bharat Refineries Limited renamed as Bharat Petroleum Corporation Limited.
	OIL spuds in Kumchai Well No. 1 on north bank of Noa Dihing
1978	OIL steps out of Assam into Orissa onshore and offshore.
	Gas supply from Bombay offshore for power generation and fertilizer manufacture starts.
1979	ONGC starts oil production in Amguri.
1980	OIL spuds in offshore Mahanadi-1 on 29 January.
	ONGC discovers oil in offshore Ratnagiri structure and gas in Dahej.
1981-82	OIL completes second Well in Mahanadi off-shore.
	Production in Assam fields slowly picks up from adverse effects of the 1980 agitation.
	SEDCO spuds first Well in Godavari offshore under contract arrangement with ONGC.
	Drilling begins in Barsilla where AOC had explored 30 years earlier.
1982-83	OIL commissions LPG plant based on the cryogenic method of manufacture. Second phase of offshore drilling begins in Mahanadi Basin. ONGC embarks on aggressive exploration policy and takes up 21 new structures for drilling.
1983-84	OIL obtains lease and finalizes exploration programme for Rajasthan. ONGC takes up 27 new structures for drilling.

Gas struck at Razole in Andhra Pradesh and Gotaru in Rajasthan. Water injection commences in Bombay High (North).

1984-85 OIL starts Vibroseis surveys in its Rajasthan concession through contractual arrangements with CGG of France. Arrangements finalized for Andaman exploration.

ONGC takes up 28 new structures for drilling.

Oil struck in Kutch offshore, Godavari off-shore and at Changmaigam in Assam.

1984-85 Drilling commences at Narimanam in the Cauvery Basin.

1985-86 OIL wins Corporate performance award given by the Harvard Business School Association of India and The Economic Times for the best public sector unit in the country.

Three significant oil finds in the south bank of the Burhi Dihing between Nahorkatiya and Moran.

First well commenced in Eastern Shelf of Andamans.

ONGC strikes oil at Narimanam thus brightening the prospects of Cauvery basin.

LPG plant at Uran commissioned.

1986-87 OIL starts onshore drilling in Mahanadi through contract arrangements with Richter Drilling of Australia.

Two more wells drilled in Andamans, one on the east and the other on the west shelf.

ONGC strikes oil in the Tapti offshore area and in the Namti structure in Assam.

Systematic development of Gandhar discovery begins.

1987-88 OIL pulls out of Andaman project temporarily.

Reassessment of prospects in Mahanadi taken up.

ONGC takes up 37 new structures for drilling.

Bombay offshore gas charged into HBJ pipeline.

EOR caustic flood pilot project started in Sanand.

1988-89 OIL establishes commercial gas in its Rajasthan concession area, and acquires license to explore over 24,900 sq. km. in Saurashtra offshore basin. ONGC discovers Nade field in Gujarat and establishes new resources nearThanjavur in the Cauvery Basin.

OIL wins Indira Gandhi Memorial National Award for excellence amongst Public Sector Undertakings.

1989-90 OIL discovers commercial oil in the Eocene of Dikom and thus opens a new and exciting horizon for exploitation. R&D complex inaugurated at Duliajan. ONGC discovers South Heera field in Bombay Offshore with approximate reserves of 13.8 MMTOE. ONGC western offshore production reaches a peak of 21.72 MMT.

1990-91 OIL discovers commercial oil in Kathaloni and drills Assam's deepest exploratory well to 5001 metres in the Talap structure. In Rajasthan, gas discovered in Dandewala structure and, more importantly, heavy oil in Cambrian sandstones at Baghewala in the Bikaner basin.

1991-92 OIL discovers commercial oil in Lower Eocene of Hatiali structure. ONGC concentrates in North Gujarat Fields.

1992-93 Gas from OIL's Tanot and Dandewala fields released for supply to Rajasthan State Electricity Board. Rectification measures well in hand to reduce gas flaring in western offshore.

1993-94 ONGC production from western offshore reaches low of 15.37 MMT Neelam development completed. ONGC enters into joint ventures for developing Ravva, Mid and South Tapti, Mukta and Panna fields.

1994-95 Spectacular come-back by ONGC in western offshore, increasing production by 5 MMT in one year to 20.23 MMT. Five important discoveries in Bombay, K-G and

Cauvery basins. OIL starts exploratory drilling in Saurashtra offshore. Oil production of 2.882 MMT becomes the highest in a decade.

1995-96 ONGC oil production of 31.64 MMT registers 8% increase over that of previous year while net gas production of 17 BCM represents 21 % increase. There were 8 important discoveries through drilling of 407 wells. Two major gas-lines commissioned in Western Region.

1997-98 ONCG crude oil production was 28.251 MMT (million toones). Oil production was about 3.0 MMT.

2001-01 ONGC crude production was 32.427 MMT. Oil production was about 3.0 MMT. Petro imports was 15664 million US dollars which was 32.3 per cent of total imports. Refinary production was 103.444 million tonnes crude thoughout.

2001-02 ONGC crude production was 32.03 MMT. Oil production was about 3.0 MMT. Petro imports was 13669 millon US dollars which was 27.0 per cent of total import. Refinery production was 107.328 million tonnes crude throughout. Imported about 75 million tonnes of crude oil. Petro products demand was 98.57 million tonnes.

2002-03 Likely to import 95.159 million tonnes of crude oil.

2006-07 Crude output will be 33.97 million tonnes. Petro products demand will be 118 million tonnes.

Appendix-E

Role of Petroleum Geochemistry

Introduction

Three disciplines, geology, geophysics, and geochemistry are beginning to be an integral part of all exploration programmes. The slowness in utilizing geochemical concepts in exploration is due to that this science itself is relatively young. Although research in subsurface petroleum geochemistry has been going on from about 1950. It is only from 1970 the geochemical techniques have been successfully applied to exploration problems on a widespread scale. During this period of gestation, geochemists had to show that these geochemical techniques really paid off in finding new oil and gas fields and avoiding barren areas.

Geochemistry was started in the research laboratory instead of in the field. There often was a lack of combination between the research geochemist and the operating geologist. To some extent, the required use of maturation techniques has bridged this gap. The paleontologist and petrographer who studies kerogen maturation can more easily be exchanged with field personnel than can the geochemist. The companies that utilize geochemistry most unsuccessfully have geochemical seminars to train new geologists and geophysicists. These companies rotate geochemists and geologists between operations and the laboratory. The sophistication of geochemical techniques requires that many of them be done in the laboratory. It is essential that the laboratory should become an open line of communications and contact with operating problems.

New and Deeper Pay Horizons

Dry holes are caused by a variety of circumstances, including sometimes simply a near miss. Oil and gas fields have been discovered by geophysicists and geologists reinterpreting old data and opening and retesting dry holes. They have also been discovered by persistent testing during driving. The geochemist can help in the decision as to whether to drill deeper or to try again by emphasizing when the hydrocarbon indicators are positive and when they are negative. There is still an enormous potential for new oil and gas fields on the continents of the world if the political, economic, and environmental problems can be reconciled. In the USA, there were about 23,000 oil and gas fields discovered as of January 1, 1968, in a sedimentary basin area of about 6.5 million km^2 or 2.5 million $miles^2$ (Halbouty 1970). This calculates to around 3500 fields per million km^2 or 9200 fields per million $miles^2$. The Russia had 1118 fields in a sedimentary basin area equivalent to 14 million km^2 or 5.4 million $miles^2$ which is about 80 fields per million km^2. Clearly, the geochemist can play an important role in minimizing the number of dry holes drilled in order to find the new oil and gas fields.

Geochemical monitoring of maturation levels is particularly important in deep wells. In 1975, 413 wells were drilled deeper than 4572 m in the USA. About 130 of these went below 5486 m. This was only 6 and 2 per cent respectively of the total 7026 wild-cats drilled. The potential for hydrocarbon is high in these intervals, because the temperatures are usually well into the oil-generating and gas-generating ranges. Oil would be found in the areas of low geothermal gradients and natural gas in areas of high gradients. More deep wells are needed, but there is the hazard that deep drilling can go beyond the greatest depths of hydrocarbon generation, accumulation and retention. Although methane (CH_4) is thermally indestructible at sedimentary basin temperaturas, it can be lost through diffusion, migration, or chemical reaction. Any well drilled beyond 6000 m should continuously monitor the organic maturation very closely. The cost of this is trivial compared to the cost of deep drilling. At present, no economic accumulation of hydrocarbons is known where the kerogen of the associated fine-grained rocks has a vitrinite reflectance above 3.5 and a hydrogen content below 3 per cent. Such a field may some day be found, but the odds are against it. The job of a geochemist is to keep the well geologist advised as to whether the odds are getting better or worse as the drill goes deeper.

Giant Oil and Gas Fields

Some subsurface conditions are such that huge source rock drainage areas contribute hydrocarbons to a large favourably situated, tightly sealed trap, and a giant oil or gas field is formed. Only a few areas in the USA probably contain giant oil or gas fields, e.g., the overthrust belt along the western Wyoming bordar, northern Alaska and the USA offshore. Few laypersons realize the enormous effect a few giant oil or gas fields have on the economics of oil. Only 8 fields contain 56 per cent of all proved recoverable oil in the world outside the Russia Block (Halbouty 1970) . Six of these fields are in the Middle East. In the Russia in 1961, only 8 per cent of the oil fields held 80 per cent of Russia's proved oil reservs. In the USA, only 46 of 23,000 fields contain 36 percent of the oil reserves. Gas is similarly distributed. Eleven Canadian giant fields produced 58 per cent of Canada's gas in 1965. Halbouty et al. (1970) defined a giant oil field as having more than 50×10^6 bbl of recoverable oil and a giant gas field as having more than 3.5 trillion ft^3 of recoverable gas. As of 1970, there were 187 giant oil fields and 79 giant gas fields. The five sedimentary basins containing the most giant fields were (1) Arabian Platform and Iranian Basin (56), (2) West Siberian Basin, Russia (29), (3) Sirta Basin, Libya (13), (4) Volga Ural Basin, Russia (10), (5) Gulf Coast Basin, USA (10).

All giant fields have in common large traps with great vertical thickness or wide lateral extent or both. Structure is not the only criteria. The presence of source rocks and the time of maximum generation and migration of hydrocarbons relative to the availability of porosity and impervious caps are equally important. Petroleum geochemists could make a significant contribution by studying the whole process of origin, migration, and accumulation for a few of these giant fields. There may be some geochemical parameters unique to the giant fields. Even the origin of the oil in that giant of giants, Ghawar (Saudi Arabia), is not clearly defined. There are no published geochemical studies of these source rocks of Ghawar field. One should study at what temperatures were these beds buried and whether the volume of potential source rocks is enough to explain the observed accumulation. Young et al. (1977) calculate the age of the Ghawar oil field as 166 million years, Middle to Upper Jurassic. If there is continuity of permeability through Jurassic carbonate sequence, the source bed drainage area may extend more than 160 km into the folded belt of the Zagros Mountains. Dickinson (1974) has postulated that partial subduction of the Arabian plate in its collision with the Eurasian plate has expedited the migration of oil along rifted-margin sediments from source beds beneath the Zagros fold-thrust belt. Certainly the high temperatures associated with the subduction zone would cause maximum generation of hydrocarbons. Halbouty et al. (1970) outline the geological factors concerning the world's giant oil and gas fields. Similar information is needed on the geochemical factors.

Continental Shelves and Beyond

Known world reserves of oil were 1032 billion bbl and gas reserves were 5,451,065 bcf (oil and gas Journal, Dec, 2001). Undiscovered resources on the continents and continental shelves are estimated to be 2700 billion bbl. At least half of this undiscovered oil and gas is believed to be on the continental shelves. Politics, economics, and environmental restrictions are three factors that could considerably reduce the quantity of world resources that will eventually become proved reserves. Even when the economics and other factors are favourable, there can be a time of lag of several decades between initial discoveries in a basin and the recognition of its true potential. The first commercial field in the Middle East was proven in 1908. During World War II Middle East oil reserves were between 5 and 10 billion bbl. In 1978 the oil and gas reserves were estimated to exceed 450 billion bbl, or 60 billion M-T (as oil). New basins are proven up more rapidly today, but still it takes 20 to 30 years before we know the full potential of areas, e.g., North Sea, Colville Basin of Alaska, Gulf of Campeche, etc.

The estimated world resources of undiscovered petroleum do not include oceanic areas beyond the continental shelves. Table No.

I lists the organic carbon content of these deep ocean sediments (Emery 1975). Abyssal plains acan be written off as having too low an organic carbon content and too thin a sedimentary section to develop the temperatures required for generating petroleum. The continental slopes, which extend to depths of 1500 to 3500 m may be suitable source beds but they appear to contain few layers of coarse-grained reservoir beds required to accumulate the oil. Coarse-grained sediments carried seaward by turbidity currents go beyond the slope to the continental rise, usually via submarine canyons. The rises are gently sloping topographic features that extend from the bottom of the slopes to depths 3500 to 5500 m. The total area of a continental rise and deep marginal basin sediments to be $62x10^6$ km^2 ($24x10^6$ $miles^2$), which is equal to the combined areas of continental selves and slopes (Emery and Skinner 1977). The rises are at least 5 km thick off eastern North and South America and Western Africa. The rises contain more than half the world's total volume of sediments. They must be considered potential major sources of oil and gas, because at least the upper sections contain coarse-grained turbidite layers interbedded with fine-grained pelagic sediments. This combination of source and reservoir facies in thick sediments with adequate organic matter is highly prospective. The average geothermal gradient of continental rise sediments to be $3.5^0C/ 100$ m. This means that at a sediment depth of 2500 m on the rise the temperaturewould be 150^0C, already into the gas generating stage. Several rise sediments contain salt domes, which can create traps and also act as a conduit for heat from deep-seated sources, e.g., Gulf of Mexico.

The rises occur where the continental margin is within a crustal plate in areas of divergence of sea floor plates. The deep marginal basins are in areas of convergence of sea floor crustal plates with continental plates. The marginal basins of economic interest are inshore of and shallower than the deep trenches along the underthrust belt. In 1976, a well was completed in 1055 m of water off Thailand when the first deep well is drilled on a continent rise or deep marginal basin, it will be most important that petroleum geochemists continuously monitor the hydrocarbons in the cores and cuttings by the best techniques available.

Important Points to Remember

1. The three important factors in a hydrocarbon survey are : (l) the quantity of organic matter – organic richness, (2) the quality of organic matter – oil or gas source, and (3) the maturation level – immature, mature, or metamorphosed. They can be determined from organic carbon analysis, pyrolysis FID, and vitrinite reflectance.

2. In rank wildcat areas, these three parameters should be determined on unweathered outcrop samples, well cuttings of former dry holes, and selected cores or cuttings from nearby areas with similar formations. Also, any bitumen samples or oil or gas seeps should be analyzed and correlated with the nearest productive area.

3. The first well in a wildcat area should have a mud logger capable of reading shale gas, or well cuttings should be taken at intervals of 15 to 30 m and canned at the well site. The initial hydrocarbon survey can be made on a small selection from this group. A temperature log should be obtained and fluid from drill-stem tests should be analyzed for traces of hydrocarbons.

4. If oil is discovered, correlation with extracts of the presumed source rocks and correlation with oils from nearby areas is important.

5. In developed areas, the shallower and deeper horizons from known pay zones as well as extensions into areas of complex tectonics should have geochemical monitoring. Variations in source facies, geothermal gradients, and tectonic history can cause wide differences in hydrocarbon prospects for different parts of a basin.

6. Significant geochemical interpretations can be made from analyzing relatively few samples in a wildcat well as long as major geochemical changes are clearly defined.

7. Published case histories of geochemical surveys involving many different tests by several independent laboratories have shown good agreement in interpretations.
8. The successful application of geochemical techniques to oil or natural gas exploration requires that the geochemist and operating geologist work in close cooperation.
9. Geochemical monitoring of maturation levels is particularly important in wells drilled deeper than 4500 m, because the drill can go beyond the maximum depth of hydrocarbon generation, accumulation, and retention.
10. There is still an enormous potential for new oil and gas fields both on the continents of the world and offshore. More than half of the petroleum yet to be discovered is believed to be beneath the world's oceans. The prime exploration areas are the continental shelves, continental rises, and deep marginal basins. The continental rises are believed to contain more than half the world's total volume of sediments. The political, economic, and engineering problems involved in drilling them will be prodigious. The exploration will required the best combined expertise of geologists, geophysicists, and geochemists to define where the oil and gas have formed and accumulated.

References

1. Dickison,W.R., 1974; Subduction and oil migration; Geology, Vol. 2, p. 421-424.
2. Emery, K.O., 1975; Fossil fuels in the ocean. In a joint meeting of National Academy of Sciences National Academy of Engineering on national materials policy, Proc; October 25-26, 1973, Washington, D.C. Washington, D.C: U.S. Government Priting Office, pp. 79-86.
3. Emery, K. O., and Skinner, B.J.,1977; Mineral deposits of the deep-ocean floor; Mar, Min., Vol. 1, Nos. 1 & 2, p. 1-71.
4. Halbouty, M.T., 1970; Geology of giant petroleum fields: Introduction in M.T.Halbouty (ed.), Geology of giant petroleum fields; AAPG

Memoir 14, Tulsa, Okla; American Association of Petroleum Geologists, pp. 1-7.

5. Halbouty, M.T., Meyerhoff, A.A., King, R.E., Dott. R.H., Sr. Klemma, H.D., and Shabad, T., 1970; World's giant oil and gas fields. In M.T, Halbouty (ed.) Geology of giant petroleum fields; AAPG Memoir 14, Tulsa, Okla; American Association of Petroleum Geologists, pp.502-528.
6. Hunt, J.M., 1979; Hydrocarbon studies in deep ocean sediments. In E.W. Baker (ed.), Symposium on organic geochemistry of Deep Sea Drilling Project Cores, Princeton, N.J., Science Press.
7. Young, A., and Mclver, R.D., 1977; Application of physical chemistry to petroleum geochemistry, Part 2: Distribution of hydrocarbons between oils and associated fine-grained sediments, AAPG Bull., Vol. 61, No. 9, pp. 1407-1436.
8. Young, A., Monaghan, P.H., and Scheisberger, R.T., 1977; Calculation of ages of hydrocarbons in oils: Physical chemistry applied to petroleum geochemistry, Part I, AAPG Bull., Vol. 61, No.4, pp. 573-600.

Table. 1

Organic Carbon Content of Deep Sea Drilling Project Samples

S. No.	*Description*	*No. of Holes*	*No. of Samples*	*Average wt. percentage organic carbon*
1.	Continental Slopes	7	239	0.50
2.	Continental Rises	18	637	0.60
3.	Closed Basins	1	58	0.58
4.	Deep Ocean Floor	158	4612	0.19

Source : Emery 1975.

Table. 2
Ranking of Oil Rich Countries — 1975

	By reserves end-1975 (10^9 bbl)	*Cumulative production (10^9bbl)*	*By 1975 production*	*10^6 b/day*
1. Saudi Arabia	(148.6)	23.1	U.S.S.R	(9.74)
2. U.S.S.R.	(80.4)		U.S.A.	(8.37)
3. Kuwait	(68.0)	16.6	Saudi Arabia	(7.00)
4. Iran	(64.5)	21.4	Iran	(5.60)
5. Iraq	(34.3)	10.5	Iraq	(2.40)
6. U.S.A.	(33.0)		Venezuela	(2.40)
7. Abu Dhabi	(29.5)	2.5	Kuwait	(1.95)
8. Libya	(26.1)	8.7	Nigeria	(1.85)
9. Nigeria	(20.2)	4.2	Abu Dhabi	(1.50)
10. China	(20)		Libya	(1.40)
11. Venezuela	(17.7)	31.7	China	(~1.3)
12. U.K.	(16.0)	0.0	Indonesia	(1.30)

(Continued from page no. 254)

	By reserves end-1975 (10^9 bbl)	*Cumulative production (10^9bbl)*	*By 1975 productiion*	*10^6 b/day*
13. Indonesia	(14.0)	6.0	Algeria	(0.94)
14. Mexico	(9.5)	5.4	Mexico	(0.71)
15. Algeria	(7.4)	3.8	U.K.	(0.04)
Total World	659			53.9

Table – 3
Ranking of Oil-Rich Contries — 1980

	By reserves end-1980	*Cumulative Production (10^9 bbl)*	*By 1980 production*	*(10^6 b/day)*
1. Saudi Arabia	(165.0)	38.8	U.S.S.R.	(12.05)
2. Kuwait	(64.9)	20.0	Saudi Arabia	(9.62)
3. U.S.S.R.	(63.0)		U.S.A.	(8.65)
4. Iran	(57.5)	29.4	Iraq	(2.60)
5. Mexico	(44.0)	7.5	China	(2.17)
6. Iraq	(30.0)	15.2	Venezuela	(2.15)
7. Abu Dhabi	(29.0)	5.8	Nigeria	(2.10)
8. U.S.A.	(26.4)		Mexico	(1.96)
9. Libya	(23.0)	13.0	Libya	(1.78)
10. China	(20.5)		U.K.	(1.60)
11. Venezuela	(18.0)	35.8	Indonesia	(1.57)

(Continued from page no. 256)

	By reserves end-1980	*Cumulative Production (10^9 bbl)*	*By 1980 production*	*(10^6 b/day)*
12. Nigeria	(16.7)	8.0	Kuwait	(1.40)
13. U.K.	(14.8)	1.7	Abu Dhabi	(1.38)
14. Indonesia	(9.5)	8.8	Iran	(1.28)
15. Algeria	(8.2)	5.6	Algeria	(1.00)
Total World	649			59.7

Table – 4

Ranking of Oil-rich Contries of Table 3 on the basis of the reserve/prodcution ratio 1980

Rank	*Country*	*Reserves/production ratio (years)*
1	Kuwait	127
2	Iran	123
3	Mexico	62
4	Abu Dhabi	58
5	Saudi Arabia	47
6	Libya	35
7	Iraq	32
	World average	30
8	China	26
9	U.K.	25

(Continued to page no. 259)

(Continued from page no. 258)

Rank	*Country*	*Reserves/production ratio (years)*
10.	Venezuela	23
11.	Algeria	22
12.	Nigeria	22
13.	Indonesia	17
14.	U.S.S.R.	14
15.	U.S.A.	8

Table – 5

Ranking of oil-rich contries — 1997

S. No.	Country	By reserves end –1998 Oil (10 bbl)	Gas (bcf)	Country	Estimated 1998 production (106 b/d)
1.	Saudi Arabia	259.00	204,000	Saudi Arabia	8.06
2.	Iraq	112.50	109,800	United States	6.37
3.	Kuwait	94.00	52,200	Russia	5.89
4.	Abu Dhabi	92.20	196,100	Iran	3.60
5.	Iran	89.70	812,300	China	3.20
6.	Venezuela	72.60	142,500	Venezula	3.11
7.	Russia	48.57	1,700,000	Norway	3.05
8.	Mexico	47.82	63,456	Mexico	3.05
9.	Libya	29.50	46,400	U.K.	2.66
10.	China	24.00	48,300	Iraq	2.11
11.	United States	22.55	167,223	Nigeria	2.08
12.	Nigeria	22.50	124,000	Canada	2.02

(Continued from page no. 260)

S. No.	Country	By reserves end –1997 (10^9 bbl)	Country	By 1997 Production (10^6 b/day)
13.	Norway	10.42	Kuwait	1.84
14.	Algeria	9.20	Libya	1.42
15.	Kazakhstan	5.42	Indonesia	1.36
	World	1019.55		65.44

Table – 6

Ranking of oil-rich contries — 1998

S. No.	Country	By reserves end –1998 Oil (10 bbl)	Gas (bcf)	Country	Estimated 1998 production (106 b/d)
1.	Saudi Arabia	259.00	204,000	Saudi Arabia	8.06
2.	Iraq	112.50	109,800	United States	6.37
3.	Kuwait	94.00	52,200	Russia	5.89
4.	Abu Dhabi	92.20	196,100	Iran	3.60
5.	Iran	89.70	812,300	China	3.20
6.	Venezuela	72.60	142,500	Venezula	3.11
7.	Russia	48.57	1,700,000	Norway	3.05
8.	Mexico	47.82	63,456	Mexico	3.05
9.	Libya	29.50	46,400	U.K.	2.66
10.	China	24.00	48,300	Iraq	2.11
11.	United States	22.55	167,223	Nigeria	2.08
12.	Nigeria	22.50	124,000	Canada	2.02

(Continued from page no. 262)

S. No.	Country	By reserves end –1998		Country	Estimated 1998 production (106 b/d)
		Oil (10 bbl)	**Gas (bcf)**		
13.	Norway	10.91	41,424	Abu Dhabi	1.89
14.	Algeria	9.20	130,300	Kuwait	1.80
15.	Brazil	7.11	8,039	Libya	2.89
	Total World	1034.67	5,144,736		66.30

Table – 7

Ranking of oil-rich contries — 2000 reported in 1997

S. No.	Country	By reserves end –2000 Oil (10^9 bbl)	Gas (bcf)	Country	Estimated 2000 production (10^6 b/d)
1.	Saudi Arabia	259.20	213,300	Saudi Arabia	7.99
2.	Iraq	112.50	109,800	Russia	6.33
3.	Kuwait	94.00	52,200	United tatesS	5.82
4.	Abu Dhabi	92.20	196,100	Iran	3.68
5.	Iran	89.70	812,300	China	3.24
6.	Venezuela	76.60	146,800	Norway	3.21
7.	Russia	48.57	1,700,000	Venezuela	3.03
8.	Libya	29.50	46,400	Mexico	3.01
9.	Mexcio	28.26	30,394	Iraq	2.57
10.	China	24.00	48,300	Canada	2.04
11.	Nigeria	22.50	124,000	Nigeria	2.03
12.	United States	21.76	167,406	Abu Dhabi	1.90

(Continued from page no. 264)

S.No.	Country	By Reserves end-2000 (10^9 bbl)	Gas (bcf)	Country	by 2000 producion (10^6 b/d)
12.	United States	21.76	167,405	Abu Dhabi	1.90
13.	Qatar	13.16	393,830	Libya	1.41
14.	Norway	9.45	44.037	Indonrsia	1.27
15.	Algeria	9.20	159,700	Brazil	1.13
	Total World	1028.46	5,278,484		67.23

Appendix-F

News in Focus

21st Century Belongs to India!

Major Achievements of Ministry of Petroleum and Natural Gas

Developing crude oil and gas resources

- 70 blocks awarded for exploration under the New Exploration Licensing Policy (NELP) in 3 years compared to 22 in previous 10 years.
- One of the world's biggest gas discoveries this year with in place reserves of 7 Trillion Cubic Feet (TCF) by Reliance in NELP-I deep sea block.
- NELP-II recorded success within 1½ years of contract signing - gas found near Surat (Gujarat) by Niko Resources of Canada.
- Oil discovery in Barmer puts Rajasthan on Hydrocarbon map.
- For the first time, 8 blocks awarded to exploit Coal Bed Methane (CBM) to enhance availability of natural gas.

  Enhanced oil security by acquisition of acreages abroad. ONGC Videsh Limited (OVL) acquired 20 per cent participation in Rosneft, Russia's Sakhalin-I offshore project with country's largest single investment of Rs. 8,000 crore for getting 2 to 4 MMT crude oil and 5 to 8 MMSCMD gas per year.
- OVL also acquires 25 per cent (3 Million Tonnes a year) crude oil share in – a producing oil field in Sudan by investing Rs. 3,500 crore.

- Vietnam Gas Field (OVL's share - 45 per cent) started production on 22.11.2002 which has 2.7 TCF in place reserve with investment of Rs. 900 crore.
- Redevelopment of Mumbai High leads to 9 per cent increase in production of crude oil and 4 per cent increase in natural gas by investing Rs. 8,100 crore in a phased manner.
- 5 per cent Ethanol-blended Petrol mandated to cover in a phased manner from 1.1.2003 in 9 States-Uttar Pradesh, Punjab, Haryana, Andhra Pradesh, Gujarat, Karnataka, Maharashtra, Tamil Nadu and Goa and 4 Union Territories - Chandigarh, Pondicherry, Daman and Diu and Dadra and Nagar Haveli.
- Royalty payable to States increased making it ad valorem - additional revenue to oil producing States, from Rs. 578 per M.ton to 20 per cent ad valorem of Rs. 1,250 per M.ton.

Towards smoke-free Kitchens

- 3.50 crore new LPG connections released within last five years against 3.37 crore in previous 40 years after independence.
- LPG now available on demand throughout the country.
- Introduction of 5 Kg LPG cylinders for the remote, rural, under privileged population.
- Deposit on LPG cylinder reduced to Rs. 650 from Rs. 900 following fall in procurement cost as Government introduced tender system.

Self-sufficiency in Refining

- Refining capacity doubled to 116.5 MMTPA achieving self-sufficiency.

Eco-friendly Measures

- Supply of unleaded petrol, low-sulphur diesel and LPG as auto fuel.
- Campaign against adulteration intensified.

Reforms

- Commendable performance by 12 oil PSUs recording profits of Rs. 12,750 crore in the year 2001-02 against paid-up capital of Rs. 5,200 crore.
- Administered Price Mechanism dismantled from 1.4.2002 to benefit consumers by fostering competition and to increase investment.
- Additional 8,659 Retail Outlets sanctioned by grant of marketing rights to M/s. Reliance Industries, Essar Oil, ONGC and Numaligarh Refinery.
- Guidelines notified on 20.11.2002 for facilitating laying of petroleum product pipelines on common carrier principle as pipelines are the preferred mode of transport with environmental benefits, cost effectiveness, etc.

  Proceeds from disinvestment in Oil Sector during 1998-2001 amount to Rs. 7,217 crore which is about 80 per cent of the disinvestment in all sectors.

Support to Martyrs' Families

- 439 Kargil martyrs' families awarded retail outlets and LPG distributorships.
- 9 retail outlets awarded to families of martyrs who successfully protected Parliament from attack by terrorists.

Gas Find Fires Energy Market

New Delhi—The huge gas reserve found in the Krishna-Godavari basin off the Vishakhapatnam coast by a consortium of Reliance Industries and Canadian firm Niko Resources, is set to change the energy map of India, sending promoters of future oil and power projects back to their drawing boards.

US-based oil consultancy firm McNaughten has certified that the reserve has the potential to supply between 40 and 55 mmscmd (million metric standard cubic metres) gas per day. The reserves have been estimated at 7 trillion cubic metres, which is near to state-owned ONGC's Vasai gas find of 9 trillion cubic metres.

The Reliance consortium expects to open the taps within two-and-a-half years. It has drilled four wells at depths of 4,000-6,000 ft.

Petroleum secretary B K Chaturvedi said the current nation wide gas demand for which allocations have been made, stands at 118 mmscmd. Against this, only 65 mmscmd of gas is available. "With demand expected to grow four times by 2025, the find will go a long way in bridging the gap," he said.

The Reliance consortium can supply the Krishna-Godavari gas right up to West Bengal in the east and Tamil Nadu in the south through GAIL's pipeline infrastructure. It can also put the gas into GAIL's arterial HBJ (Hazira-Bijaipur-Jagdishpur) pipeline in Madhya Pradesh through its proposed pipeline to feed the big northern market.

Reliance already has permission to build a gas pipeline from its 27 mmtpa refinery site in Gujarat's Jamnagar to Goa through Madhya Pradesh and Hyderabad. GAIL has some pipeline network in Bengal and Bihar and is working on a Vijaywada-Hyderabad-Secunderabad southern network.

With gas coming from the comparatively smaller reserve found by UK major Cairns in the same area, the system will be nearly brimming with gas in a few years.

The first casualty is likely to be US major Unocal's plan to sell piped Bangladesh gas, hanging fire due to political opposition being faced by Dhaka.

A hint of this came from oil minister Ram Naik. "With the Reliance find and ONGC-Videsh's acquisitions in Sakhalin, Vietnam and now Sudan, India has moved closer to securing its energy needs after deregulation," he said. The country at present imports 70 per cent of its crude requirement worth Rs. 80,000 crore a year.

The new find also puts a question mark over the upcoming LNG (liquefied natural gas) projects, both in the east (Kakinada, Ennore and Kochi) and the west (Dahej) coast. Shell, British Gas and Petronet LNG, which is a consortium of Indian Oil, ONGC, BPCL and GAIL.

ONGC Soon to Become a Total Oil Co.

State-owned exploration major Oil and Natural Gas Corporation (ONGC) is a step away from becoming the country's first fully integrated oil company The Public Investment Board (PIB) has cleared its acquisition of Mangalore Refinery and Petrochemicals Ltd (MRPL) for Rs 59.43 crore.

The deal now needs the approval of the Cabinet Committee on Economic Affairs (CCEA), which is a mere formality.

The ONGC is paying Rs. 2 per share on a face value of Rs. 10, totalling Rs. 59.43 crore ($12.2 million).

It wants to acquire 13.62 per cent more through a debt-restructuring package agreed with the institutional lenders. The acquisition, alongwith the debt-restructuring, is likely to be completed by March 31.

The MRPL is a joint venture between the Hindustan Petroleum Corporation Ltd (HPCL) and the Aditya Birla Group, with each holding 37.4 per cent shares of the company. The remaining 25.2 per cent is with the public.

After restructuring, HPCL's stake will drop to 16.9 per cent and the institutional lenders will have about 20.8 per cent, while the public holding will come down to 11.3 per cent.

The finance and petroleum ministries had differed on whether ONGC's buyout of the A V Birla group required PIB approval. The petroleum ministry was of the view that PIB approval was not required as the HPCL had already taken its sanction while forming the joint venture with the A V Birla group.

The ministry had said that this de facto gave clearance to any other public enterprise to invest in the venture. Besides, the PIB route would be time-consuming and would hold up ONGC's investment plans.

As per the norms of the department of public enterprises, any capital expenditure exceeding Rs. 200 crore in a joint venture, to be established by a state-owned entity, needs approval from the PIB.

The ONGC, already, has permission to set up 600 retail outlets, particularly in southern region. It has also taken equity in the Mangalore-Hassan-Bangalore product pipeline that will make it easier to push petro-products into the interiors.

With the PIB approval, one can expect to see some changes in MRPL's business. Crucial among these will be a decision to review the refinery's three-year-old crude supply agreement with Chevron-Texaco.

ONGC's logic is to reduce the proportion of imported crude oil vis-a-vis domestic crude, as it will help bring down the refinery's losses. In 2001-02, the MRPL incurred a net loss of Rs. 492.47 crore, taking the total losses last fiscal to Rs. 799.14 crore.

MRPL's agreement with Chevron-Texaco ended on December, 2002, but was extended for a month. Once ONGC takes a majority stake, it intends to feed MRPL with crude from its onshore fields in Gujarat. Driven by economics, the ONGC may divert its entire 3 million tonne per annum (mtpa) Gujarat output to the MRPL and not to Indian Oil's Koyali refinery.

MRPL has an economic advantage in Karnataka and Goa as well as south Maharashtra and north Kerala. After ONGC's takeover, the MRPL will be the only Indian refinery operating with crude produced by itself.

ONGC's current annual production of crude oil is 24 mtpa. MRPL's installed capacity is 10 mtpa, but last year the government let it produce only 4.9 mtpa. The MRPL plans to hike its production to utilise the installed capacity better.

Bigger Gas Finds Lie on East Coast

The east coast will be the country's future oil and gas bowl, with a potential of supplying up to seven billion tonnes of oil and oil equivalent gas, the country's caretaker of hydrocarbons assets has said.

"The big gas finds reported by Reliance and Cairn off the Andhra coast have already borne this out. This is just the beginning," director-general of hydrocarbons Avinash Chandra told Times News Network.

The two gas finds make up about half a billion tonnes of oil equivalent gas and at a pessimistic estimate, 6.5 billion tonnes more is waiting to be tapped, Chandra said.

Reliance and Cairn have reported seven tcf (trillion cubic feet) and one tcf gas finds, respectively, from two blocks in the Krishna-Godavari basin which were auctioned in 1999 by the government as part of a policy to bring private investment in exploration. The Oil and Natural Gas Corporation (ONGC) too is working in an adjacent block and has said it is near striking gas.

Besides the Andhra coast, areas south of Rameswaram and southwest of Thiruvananthapuram have been identified as having huge potential. The Orissa coast too has thrown up a big potential.

The cost of identifying the blocks? Roughly Rs 18 crore. "Starting in 1995, we enlisted the services of European and US satellites that passed over the areas 20 times a year to identify the potential blocks deep under the seabed," Chandra said.

Chandra said the existing geographical and seismic data collected over the years by ONGC and Oil India Ltd showed that the seabed in the east was similar to the oil and gas producing areas off the coast of Brazil. "We needed a new approach and to go deep under the sea," he said. The satellite mapping was followed with a two-dimensional seismic survey done from specialised ships.

Chandra said the east coast potential will allow India to strike harder bargains while importing gas and oil as it will narrow the gap between supply and demand for fuels. India imports 70 per cent of its requirement of 105 million tonnes of crude oil and has 65 mmcmd (million cubic metres per day) of gas against a demand of 150 mmcmd.

The government offered 100 blocks identified by the DGH in three rounds of auctioning since 1999. At least 68 of these were awarded after bids were received for 73 blocks, Chandra said.

Against this, 72 blocks were offered before the auctioning, bids were received for 13 blocks and only five were awarded.

Comment: Energy superpower in the making or just gas? Seriously, this has stupendous implications for the country, from national security to the import bill to retail prices. Hopefully, this will be the spark that lights India's future.

ONGC to invest Rs. 985 crore for Bassein East

ONGC will invest Rs 985 crore to develop the Bassein East field, which is estimated to contain 97 million tonnes of oil and oil equivalent gas. "We plan to bring to production the Bassein East oil and gas field in 30 months," ONGC chairman and managing director Subir Raha said. The field is estimated to yield 6.185 million tonnes of oil and 3.175 billion cubic metres of natural gas over 2006 to 2021, he said.

Oil reserves

ONGC said on Thursday that it has discovered oil reserves off the Mumbai coast. The Vasai West is estimated to contain 240 million barrels of in-place oil plus oil-equivalent gas.

ONGC Ties Up with Russian Firm

ONGC has tied up with Russian firm Zarubezneft for technical know-how to increase crude production from its fields in the northeast, including Assam.

ONGC chairman Subir Raha said ONGC would discuss a turnkey project wherein the Russian company would bring in their own equipment and rigs needed for raising the output. The turnkey plan would save a lot of cost to ONGC, he said, adding that the deal, if happens, would be worth more than the present Rs 12.5 lakh technology tie-up.

Raha said ONGC was looking at raising crude production from its Assam fields to 2 million tonne per annum from 1.85 million tonne a year at present. Its sister concern, Oil India Ltd, produces 3 million tonne of crude every year from its fields in the northeast.

Explaining the choice of the Russian firm, Raha said their technology and cost best suited ONGC. Assam rocks are tight formations that need specialised technique.

Up May Hold Key for Oil Hungry Nation

There is a possibility of striking liquid gold in UP. For the past year Oil and Natural Gas Corporation (ONGC) has been evaluating the hydrocrabon potential in the Gangetic plains and has launched exploration in two areas – Azamgarh and Gandhak region and Bareilly and Shahjahanpur.

The last round of investigations in 1960s and 1970s were discouraging and sent them back to the drawing fill board.

However, enhanced technology and knowledge, I through data interpretation of similar areas through surface geological mapping, has revealed the presence of hydrocarbons in some of the rocks in the Gangetic plains.

Developing countries with high energy requirements India and China are regarded as Oil Hungry nations. Indian requirement of oil is 100 million tonnes per year which will increase by 1.5 to 3 times the current consumption by 2010.

According to Director (Exploration), ONGC, Y B Sinha, at the given rate of exploitation the .7 billion tonnes of oil reserves will last up to 20 to 22 years. Currently, India is producing 30 million tonnes of oil every year and 25 billion cubic metres of gas".

To avoid developmental activities coming to a standstill under the New Oil Exploration policy there has been a renewed vigour to find new areas for oil, which is why there is an emphasis on deep sea drilling.

India spends Rs 70,000 crore on oil imports. With a decline in production in one source, new places need to be prospected. The only problem is that in the Gangetic plains there is a strange predicament.

Areas closer to the Himalayan range, due to thicker sedimentary deposition and the basin dipping towards the Himalayas, may result in hydrocarbon formation in the northern regions of the Gangetic plains.

Over a Barrel : Perils of Privatising Oil Sector

Oil and steel form the spine of the Indian economy. While privatising the steel sector may be justified because the country has emerged as the eighth largest producer of steel, there is no justification for privatising the oil sector. India imports about 70 per cent of its oil requirements which costs the exchequer Rs 86,000 crore per annum. India's oil needs are about 1.65 million barrels per day (bpd).

After nationalising the oil industry in the 1970s, India's oil imports continued to spiral which, in turn, affected industrial and economic growth. This inflated the price of building new tankers which carry crude oil and naphtha into India. Hence, India can never produce enough oil to satisfy domestic needs as onshore reserves, with some exceptions, have been exploited. Recognising this harsh reality, India has ventured into deep sea drilling to meet domestic needs and has now signed a $4.5 billion pact with Iran for supplying oil and gas. A new gas pipeline between Iran and India is on the anvil with both countries examining the feasibility of laying it over land or below the sea.

Now, only oil reserves in the deep sea basins are available for exploitation. The government has offered these to the private sector as part of the New Exploration and Licensing Policy (NELP). But drilling for oil in the deep sea is very different from working onshore. Deep sea oil reserves have to be extracted using more expensive technology than that used for onshore drilling.

During the 1971 war against Pakistan, the foreign-owned private oil companies refused to supply fuel to the Indian Navy and the Indian Air Force, prompting Indira Gandhi to nationalise the oil industry so that the government controlled prices in this crucial sector. The government did this by introducing its own machinery called the administered price mechanism. After nationalisation, the oil majors such as Indian Oil Corporation (IOC), Hindustan Petroleum Corporation Ltd (HPCL) and Bharat Petroleum Corporation Ltd (BPCL) had to keep a minimum stock level of oil to be supplied to the military when needed. These oil reserves were called the military stock levels (MSLs).

But with privatisation, all this will change because private players will focus only on profits with requirements such as MSLs taking a back seat. This may undo all that was done by Indira Gandhi in the 1970s with the result that even the security of the country may be jeopardised. Oil prices will be dictated by free play of market forces with the domestic consumer left to their mercy. After Independence, India got technical know-how from the former Soviet Union which alone sent its geologists to identify oil-bearing sedimentary strata and train Indian geologists in oil prospecting. But later the two partners fell out and the Soviet geologists were sent packing with the Oil and Natural Gas Commission (ONGC) as it was then known, taking up the onerous task of exploration and production.

Today, while drilling and production is done by ONGC, the oil majors like IOC, HPCL and BPCL refine, market and distribute petroleum and its products. All these companies are profit-making entities which contribute a sizable chunk to the national exchequer by way of taxes. Privatising these oil majors may lead to a reduction in their tax disbursements. On the other hand, the technology for steel production has improved in leaps and bounds so that today India's steel is comparable in quality with that of South Korea and Japan while this has not been the case in oil production. It is true the ONGC has entered into joint ventures for exploration and production of crude in countries such as Sudan. But it has grown in a largely monopolistic environment because there are no other major players in exploration and production who can threaten its huge edifice.

It is important to observe here that the government still controls steel distribution to crucial sectors like the military, railways and power sectors. These areas are far too sensitive to be left to the private sector. So, it is a paradox that the government chooses to hand over the entire oil industry to the private sector by disinvesting its equity in the oil majors. Disinvestment is a result of a cash-strapped government desperately hunting for funds. That is acceptable. What is not is the fact that the oil industry is arguably the only one in India which has both direct and indirect effects on the military, all other industries and by that yardstick on economic and industrial growth.

The government should have set up an expert committee to study the effect on handing over the entire oil sector to foreign majors who will only use the existing infrastructure to rake in more profits to recover their costs. It is true that committees have been set up in the past but the effect on the steel, power, military, insurance and banking sectors after privatising the oil industry has not been studied in-depth.

The Indian National Congress at its 1904 meeting declared that the country could attain economic growth only after Independence. It is true that in some sectors like steel, self-sufficiency has been reached, but oil production is a negation of this thinking. Handing over India's oil majors to global giants to raise resources will negate the concept of *swarajya* and, in effect, means handing over the economy to the foreigners whom we threw out 56 years ago.

ONGC Takes a Deep Plunge for Oil Contract

The state-run Oil and Natural Gas Corporation (ONGC) on Sunday' announced a Rs.3,600-crore exploration contract – the single largest service contracts by any corporate entity in the country – to hunt for oil and gas at depths of 3,000 metres under the seabed.

The contracts were awarded to the US-based Transocean and Dolphin for hiring deep-sea drilling and oil-producing equipment under the company's strategy to earn Rs.6,44,000 crore, which is equivalent to 10 Union budgets, in the next 25-30 years.

Code named 'Sagar Samriddhi' the project envisages investing $ 2.6 million per day in search of one-third of the estimated 11 billion tonne of oil and gas lying in deep waters, said ONGC chairman Subir Raha.

The Rs.3,600 crore will be spent over three years. ONGC will pay $ 3,61,762 a day to Dolphin, while Transocean will get $ 3,36,906.10 a day The fees will be inclusive of integrated management services that Dolphin and Transocean will provide in consortia with Tidewater and Schlumberger respectively.

The bidding process had seen some controversy, with one single agent representing three bidders. The Delhi-based Enpro Group promoted by Hari and Shyam Bhartiya represented two bids of Transocean Holdings and one by Transocean Offshore for the category of 1,800 metres and above.

Asked whether this contravened corporate ethics or vigilance norms, Raha said: "This is a business decision made in total transparency and complying with all legal and ethical parameters. We are clear in our mind and will hold our own before any constitutional body," he said.

Sources said ONGC had brought the Central Vigilance Commission (CVC) into the loop after the fact of a single agent representing three bidders came to light.

Transocean had initially asked for $ 3,51,607.20 dollars a day and Dolphin $ 3,62,877.37, The post-bid negotiations saw Transocean lowering its rates four times, while Dolphin lowered its rates just once.

ONGC director (exploration) Y B Sinha said with the new oil hunt, the firm's current annual crude oil production of 26 million tonne can go up to 49 million tonne by 2011-12 and 62 million tonne by 2016-17 for the discoveries anticipated from the campaign.

ONGC Looking for Partners for Deep Water Drilling

Oil and Natural Gas Corporation (ONGC) is looking for partners for its high-profile deep water drilling programme 'Sagar Samriddhi' launched last month.

"Contracts worth nearly one billion dollars during the next three years are being offered for the biggest ever deep water exploration effort, taken up anywhere globally, for which we welcome partners," ONGC chairman and managing director Subir Raha said on Monday.

With the administered price mechanism in place, he was confident that more foreign players would come into India.

ONGC, he said, had recently achieved zero flaring on an offshore platform and would like to do so at other platforms also.

A leading Singapore-based weekly on petroleum and natural gas has ranked ONGC first in profits as an exploration and production company for the calender year 2001, he announced at the India Oil & Gas Review Symposium and International Exhibition.

Platts', the magazine, has ranked ONGC eighth in assets and seventh in revenues while as an energy company, it stood at 42nd position as against 64th earlier, 109th in assets, 15th in revenues and 25th in profits, Raha said.

Petroleum and natural gas joint secretary, J M Mauskar, said with barriers being removed in phases, improved technologies and service providers should be able to partner with ONGC to make it a world class company

Mauskar said a record number of eight seismic vessels have been deployed and investments in the coming years is expected to be in the region of $5.7 billion even as gas hydrates have shown rich promises.

The deputy minister for department of energy of Alberta (Canada), Ken Smith, offered his country's expertise in the energy sector.

He said Alberta was a hot spot in the energy sector supplying 13 per cent of gas and 11 per cent of oil to the US. Between 1996-2001, $17 billion had been invested in this sector, bringing in $7 billion income by way of royalty last year, he said.

IOC, ONGC to Set Up Trading Desks

As part of its efforts to make Indian firms globally competitive, the government has allowed refiner Indian Oil Corporation (IOC) and exploration firm Oil and Natural Gas Corporation (ONGC) to set up oil trading and risk management desks.

"IOC and ONGC have been permitted to set up specialised trading desks to further optimise oil procurement by tapping market opportunities and enhancing use of risk management tools to mitigate price risks arising from volatile international oil prices," government sources said.

ONGC plans to set up a trading desk in New Delhi for optimising procurement of crude oil for its subsidiary Mangalore Refinery and Petrochemicals Ltd (MRPL), marketing crude oil received from investments in oil fields abroad and exporting products.

IOC, India's largest petroleum refiner, wants to scrap its current system of importing crude oil through tenders and adopt modern practices.

The two companies are likely to appoint consultants for setting up the trading and risk managing desks shortly, sources said.

In a related development, the government has allowed ONGC to procure crude oil for its subsidiary MRPL directly.

"With this, ONGC would import all crude including the term contract with Iran, for MRPL," they said.

Sources said IOC was studying a report of KPMG, the consultant it appointed, to suggest options for setting up oil trading desk in India. IOC also plans to operate in the Singapore over-the-counter market for risk management.

The company, which buys almost half of its 32 million tonnes of crude oil imports from the spot market, had been talking to British Petroleum for a year on cooperating in crude imports and risk management. Last year, it abandoned talks and decided to go it alone.

IOC chose India over London as it provided the advantage of operating in both the Asian and the European time zone. A trading desk in India would be able to capture the Singapore market when it opens in the morning and the American market when it closes.

ONGC wants to optimise export of surplus products like jet fuel, gas oil and fuel oil from MRPL as also its own production of value added products like LPG, kerosene and naphtha, which till now are being done through tenders that give little margins, sources said.

Besides managing risk and hedging volatalities in the international oil market, the desk would also help the firm locate buyers for 3 million tonnes per annum crude oil it would receive from a Sudan oil field.

ONGC is bringing 1.3 million tonnes of Sudan crude for processing at MRPL this year and has hired Trafigura of London to trade its remaining share in the Greater Nile Oil Project in Sudan.

IOC Eyes Lanka and Bangladesh

State-owned Indian Oil Corporation (IOC) is scouting opportunities in other countries, including Bangladesh and Indonesia, and would invest about Rs 7,000 crore in a naphtha cracker plant.

"We are looking for opportunities in a host of countries, including Bangladesh, Sri Lanka and Indonesia," IOC chairman M S Ramchandran told reporters after the company's annual general meeting in Mumbai.

IOC is looking for securing equity oil and gas – independently or with partners – from Iran, Kuwait, Sudan, Bangladesh, Qatar, Papua New Guinea, Myanmar and Indonesia, he said, adding that alongwith ONGC, it has been made non-operators for the northern Kuwait project. IOC was looking at an investment of Rs.7,000 crore for setting up a naphtha cracker plant, he said and added that Rs.5,000 crore would be spent for setting up PXPTA plant and Rs.3,000 crore for lab.

ONGC Plan Rs. 6,000 crore Overseas Investment

Oil and Natural Gas Corp said it will invest over Rs. 6000 crore annually in acquiring oil properties abroad and is currently in dialogue for buying out stakes in oil fields in more than a dozen countries including Australia, Indonesia and Russia. "We have budgeted more than Rs 6,000 crore annually for acquisitions of oil and gas fields abroad. This figure may go up depending upon the investment requirement for developing a particular field," ONGC chairman and managing director Subir Raha said. ONGC's overseas arm OVL, has to date taken participating interests in oil and gas field in nine countries including Vietnam, Russia, Libya, Syria, Iran, Iraq, Sudan, US and Myanmar.

Cairn Energy Strikes Oil, Gas in Rajasthan

Britain's Cairn Energy Plc has struck oil and gas in Rajasthan, the fourth discovery by the British firm in the block which borders Gujarat. "We have encountered hydrocarbons RJ-Q-1 exploration well in block RJ-on-90/1 in Rajasthan," Cairn Energy chief executive Bill Gammell said. Previously, Cairn had found oil at Guda and Saraswati and oil and gas at Raageshwari in the same block, totalling over 30 million tonne of oil reserves.